Raising Rabbits

Raising Rabbits

by

Ann Kanable

 Rodale Press Emmaus PA

Printed in the United States of America on recycled paper

Library of Congress Cataloging in Publication Data

Kanable, Ann.
   Raising rabbits.

   Includes index.
   1. Rabbits. I. Title.
SF453.K25   636'.93'22   77-23926
ISBN 0-87857-183-3 hardcover
ISBN 0-87857-314-3 paperback

6 8 10 9 7   hardcover
12 14 16 18 20 19 17 15 13 11   paperback

Dedication: *To E. J. Duncan.*
He helped so much.

Contents

Introduction

For the person wanting to produce all or most of his own food, domestic rabbits can be most valuable livestock. They can be raised in backyard hutches in certain urban areas, on large farms, on a small place in the country, almost anywhere. Rabbits are one of the least expensive meat animals to buy when the potential breeder is shopping for foundation stock. On even the most leisurely of breeding programs, a good rabbit doe can produce from three to five litters per year, numbering between six and twelve young per litter.

Many does conceive and kindle up to seven litters per year on a commercial breeding schedule. This requires special creep feeding of the young rabbits after weaning. The doe is rebred when the little rabbits are only about two to three weeks old. Young are weaned at four to five weeks of age, when they are not yet ready internally for a full diet of the regular rabbit feeds.

Stepped-up commercial breeding schedules are not necessary for putting good meat on the table, and they run into too much work.

For the homesteader, the most practical breeding program is one of the slower ones still in use in commercial rabbitries. The doe is returned to the buck when the litter is from six to eight weeks old. The litter is left with the doe for another two weeks, the little bunnies are weaned. By this time, the bunnies will already be eating a great deal of the pelleted feed or grains, and can be given some of whatever other foods the mature rabbits get.

Rebreeding the doe six to eight weeks after kindling will give the homesteader five to six litters of fryers per doe per year. In most years, there will be five litters. Occasionally, a sixth litter is obtained when the doe's last previous litter was kindled very early in the fall.

Contrary to what some rabbit people say, the best results are obtained from a rabbit herd when the does are kept working on a fairly regular schedule year-round. The reason: idle rabbits become fat; fat rabbits become lazy, not wanting to breed at all; and fat, lazy idle rabbits become sterile. This sterility can become permanent in time, causing a total lack of production in the herd.

Precautions will have to be taken by the herd manager during very hot weather to protect the does from having heat stroke while kindling. During extremes of cold, measures must be taken to protect all rabbits, but especially the newborn babies, from becoming chilled. Rabbit babies are born completely naked and chill easily in frigid weather. These difficulties can be overcome, though, with just a little foresight on the part of the breeder.

One of the major things most experienced rabbit people will warn the beginner is: "Watch out for the 'expert', the person who knows all there is to know about rabbit raising and then some." If you're confronted by one of these experts, run for your rabbits' lives!

The person who has had much success with rabbit breeding is more likely to admit that he really doesn't know a lot about rabbits. If you ask him for help with a specific problem, he is apt to tell you, "I don't know how well it would work for anyone else, but this or that worked well for me in my herd." He knows that no two rabbit herds are exactly alike, and no single remedy will work exactly the same in two different herds.

Veterinary researchers are among those who readily admit that not enough is understood about domestic rabbit husbandry. No one is yet able to explain precisely why two herds, suffering from the same apparent problem, will react differently to identical corrective measures. It is believed that circumstances in a given herd play at least some role in this (climate in a certain locality, atmospheric pressures, management practices, even the rabbitry operators' individual attitudes toward the animals they tend are suspected to have some effect on the reaction of the herds.)

There are virtually no remedies for ailments that will work equally well in each and every herd. Therefore, what a rabbit book must do is try to set forth some guidelines that will enable the individual rabbit raiser to work out methods best suited to his own locale, weather conditions, and his own herd's physical requirements.

To begin with, most people buy more rabbits than they actually need to put meat on the table. They wind up eating rabbit so often, they begin to wish that some of their does would miss just once in a while! The average family of two adults and two children, having access to other types of meats, needs no more than six does, even if they like rabbit meat well enough to eat it more than once a week, year in and year out.

If six good does kindle five times each year, and raise out just seven fryers to the litter, that amounts to 210 fryers per year for the table. At a 50 percent average dressout, that adds up to somewhere between 425 and 525 pounds of good white meat.

Of course, there is always the chance that a doe will fail to conceive, or will lose an entire litter for some reason. It happens in the very finest of rabbitries, both commercial and show, and I'm sure it happens to other homestead herds too.

The first really bad mistake beginners make is that of buying the rabbits first, having no housing ready for them. Second, many people buy more rabbits than they can handle easily, having no prior knowledge of the many small jobs that must be done regularly to insure good health in the herd. The third serious mistake is thinking that rabbits can make do with just any cage—rabbits don't need much space, anyway. A fourth mistake is in thinking that rabbits can thrive on a little bit of greens from the garden, and some water whenever one happens to think about it. T'ain't so, believe me!

Of all the types of livestock useful on the homestead, with the possible exception of the young turkey, the rabbit is the most easily stressed. While domestic rabbits are relatively free of disease when kept in clean, roomy quarters with water and nutritional needs met, they do react rather quickly to stressful situations. A lack of water, poor nutrition, cramped and crowded quarters, dirty conditions— these are only a few of the things that can wreak havoc in rabbit production.

At the start, not only from the work management angle but from a financial view as well, the beginner should not buy more than two or three does and just one buck. Once you learn all of the little jobs that are essential to the health and top production of your herd, you can quickly increase herd numbers by saving young does and bucks from the best original stock. Starting with just a few rabbits also gives the beginning breeder time to learn if you really *like* working with rabbits. Being stuck with animals you don't like and detest caring for can become a real pain in the neck, and the rabbits soon sense the handler's antagonism. Production is slowed down or even stopped altogether, and the raiser is stuck with unproductive, expensive-to-feed animals.

Now, it's true that rabbits love greens and root vegetables from the garden. Ask any gardener who has an unfenced vegetable plot! But rabbits need more than a few greens or carrots to maintain good health and production. They need grains, dried grain stalks such as oat or wheat straw or hay, alfalfa hay, and so on.

My husband and I always preferred to base our rabbits' diet on a good brand of pelleted feed, supplementing this with such things as freshly pulled clover, dandelion greens, carrots with tops still on, plantain weed leaves, and just occasionally sassafras leaves and small pieces of the sassafras limbs with the bark left on, good clean oat hay or straw, or a piece of fruit or fruit peel. (Surprisingly, some rabbits are wild about grapefruit pulp and peel!)

By using some of the pelleted feed, rabbit fryers seem to finish out (reach slaughter weight) more quickly, with more tender, more succulent meat. The longer the rabbit takes to reach slaughter weight, the tougher the meat will be.

Rabbits can be raised quite successfully without using pelleted feeds, but you must have a good understanding of the nourishment afforded by various types of grains to accomplish this. No one grain can do the job. It takes a combination of oats, corn, wheat, barley, roughage (in the form of hay, as a rule), along with other foods to provide adequate vitamins, trace minerals, and other nutrients. Thus, it is much easier, for the average person, to make use of some of the pelleted feeds available and add supplementary foods to make sure the rabbits are getting ample supplies of all the nutrients they need. The pelleted feeds are rarely sufficiently fortified to meet the total needs of the active, healthy rabbit.

Rabbits, like any other animal, have different personalities. Sooner or later, virtually every beginning rabbit raiser comes face to face with a rabbit that has an absolutely filthy disposition. It may be a doe, a buck, or even a very young rabbit. The most common displays of bad temper come from does with brand-new babies in the nest. These flurries do not necessarily mean that the doe is just naturally mean. Her ill-temper may be no more than the doe's natural instinct to try to protect her young from anything that appears to be an outside threat.

Most does can be handled quite successfully when they have new babies in the nest by simply giving them a small amount of attention for themselves, before trying to handle the young. Sometimes, giving the doe a special tidbit to eat will keep her occupied while you look over the litter. Does are usually very hungry after kindling and starting to make milk, and an especially tasty morsel, such as a carrot or a bit of fresh clover takes her mind off the nest for a while. Many times, just rubbing the hand back over the head and ears of the doe and talking to her quietly will calm her fears that you intend her little ones some harm. This works especially well if the does are accustomed to being touched and hearing the breeder talk during regular chores.

It isn't likely that anyone would deliberately keep a truly vicious dog, swine, or bull around to endanger humans. To me, it is equally foolish to keep a really nasty tempered rabbit. This is doubly true where there are small children who may reach into a doe's cage. Take it from one who has been very painfully bitten and clawed on a few occasions, the rabbit may be smaller than the dog, pig, or bull, but they can still dish out considerable damage. A mean rabbit could be especially damaging to the small curious child. Those claws and teeth are razor sharp, and could easily blind a child, or leave ugly, ragged scars on the child's face. It is so senseless to keep a vicious rabbit, when the majority of these animals are rather gentle, friendly creatures. The proper place for the perpetually angry rabbit is the family dinner pot!

Chapter 1

Housing and Equipment

The first mistake, and one of the worst, the beginning rabbit breeder can make is in buying rabbits when he doesn't have the housing ready and waiting for them. This necessitates the use of temporary quarters. All too often, the temporary accomodations end up becoming the permanent housing, and the rabbits wind up living in poorly constructed quarters, with damp, drafty winds whistling through in cold weather.

In recent years, the trend in commercial rabbit housing has been toward total environmental control in large, heavily insulated buildings erected especially for the production of meat fryers. The animals never see natural light. Whatever fresh air they receive is pumped in by fans, which must be operated day and night. The rabbits usually need vitamins, which are administered through the drinking water, as a rule. Since all outside natural light is shut out, electric lights must be burned whenever light is needed.

These methods are terribly expensive and not very feasible for the small raiser who is interested almost exclusively in producing meat for his own use. Therefore, we will concentrate on those methods of housing rabbits that can be erected at a more reasonable cost, and usually by the homesteader himself.

My husband and I have raised rabbits in outdoor hutches, in open, pole-type sheds, in frame buildings with the upper halves of walls made of one-inch poultry screening, and in an all-aluminum barn, also with the upper halves of sidewalls covered with the poultry mesh. It was our experience that fewer problems with the rabbits, healthwise, were encountered in the outdoor hutches and in the open, pole-type shed. Flies were rarely a problem around the outdoor hutches or in the shed. The small, frame barn had only a slightly worse fly population. When we transferred the major portion of our herd to a new, all-aluminum barn, we began noticing a definite increase in the number of flies, even though sanitation methods were the same for all methods of rabbit housing.

Due to the fresh air circulation and drainage, moisture was never a problem except in the metal barn. The dirt floor of the open shed was always dry and odor-free. There was also an earthen floor in the metal barn, but due to the interference with free air flow by the lower solid walls of metal, fresh air couldn't dry out those spots that became soaked with urine.

Another interesting thing, which our own experiences verified, we learned from the men who picked up our meat stock for a processor of rabbit meat. These rabbit haulers told us, more than once, that the very healthiest rabbits they handled came out of outdoor hutches. Their next choice of rabbit housing was the open pole shed, then the pole barn. Last choice of rabbitry housing among these professional haulers was the building with totally controlled environment! It was their belief that the environmentally controlled building "was built for the comfort and convenience of the rabbitry operator more than for the benefit of the rabbits."

Therefore, it is our firm belief that for the small breeder, whether homesteader or otherwise, the outdoor hutch is the method of producing the healthiest rabbits possible. Of course, it is sometimes a little less convenient to tend the rabbits. When the weather is really terrible, it is much more comfortable to do the feeding chores inside a nice cozy building.

Outdoor Hutches

Probably the most practical type of housing for the average small raiser or homesteader is the outdoor hutch. If one

This sort of jerry-rigged housing often suits the rabbits just fine, but it looks like the devil and can turn close neighbors against your rabbit raising.

lives where there is no danger of complaint from neighbors, rabbits can be kept in outdoor hutches. Living in town can present some problems, especially if hutches and the immediate surroundings are not kept absolutely spotless and measures taken to keep flies away.

Many people do raise rabbits in outdoor hutches in urban areas. By using extra care in sanitation, by erecting some sort of attractive screening device to camouflage the hutches, and by minimizing the fly population that is attracted to any sort of animal enclosure, they raise a part of their own meat in areas which normally discourage any sort of home production of meat animals.

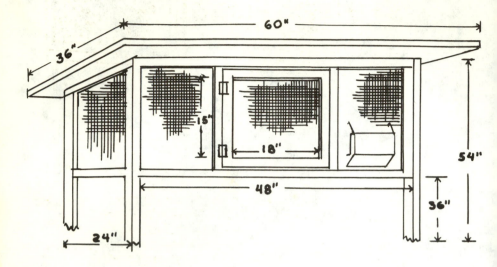

A serviceable, durable, and relatively inexpensive
outdoor hutch can be made using a wooden frame-
work and rabbit-cage wire. The general dimensions
are shown. The hutch can vary from eighteen to
thirty inches wide and from thirty-six to forty-eight
inches long. Relate the size of the hutch to the length
of the rabbit keeper's arm: he or she must be able to
reach easily into all corners of the cage. The main
frame should be made of two-by-fours, the door
frame one-by-twos. The roof can be a sheet of plywood covered with ordinary shingles or roll roofing.
It should overhang the hutch by at least six inches
on each side.

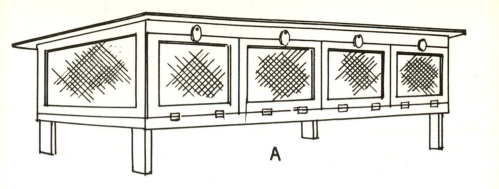

A

Here's a different approach to constructing an outdoor hutch using wood for framing and rabbit wire for caging. The basic unit (A) provides four cages, but this unit can be doubled to provide eight cages or stacked in single or double arrangements to provide eight, sixteen, or even more cages (B). The basic di-

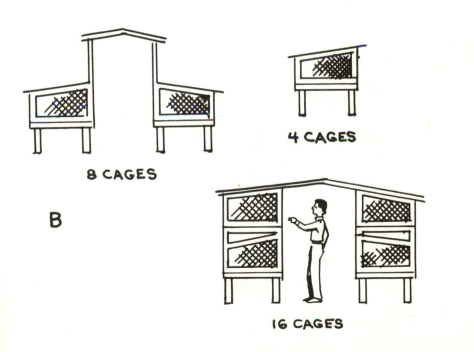

8 CAGES

4 CAGES

B

16 CAGES

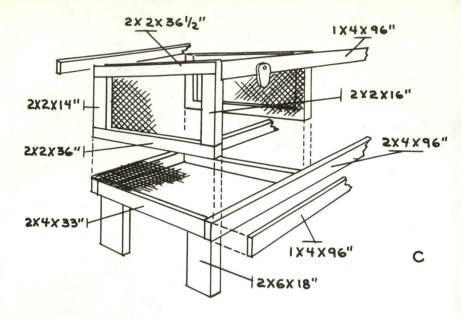

2X 2X 36½"

1X4X96"

2X2X14"

2X2X16"

2X2X36"

2X4X96"

2X4X33"

1X4X96"

2X6X18"

C

mensions and materials are shown. The base is a two-by-four frame with four two-by-six legs. It is covered with rabbit wire. Then the five identical end wall/partition assemblies are constructed of two-by-two, covered with rabbit wire, and attached to the base. The one-by-four fascia boards are added as in-

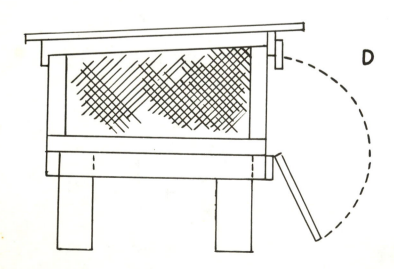

D

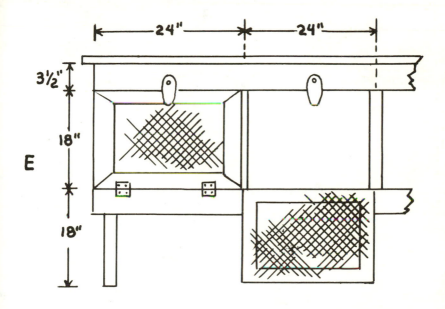

dicated front and back. The back is closed in with
more rabbit wire, and doors, framed of one-by-three
and covered with wire, are installed. The roof can be
plywood or boards covered with shingles or roll roofing. In extremely cold weather, the hutch can be
temporarily closed in with plastic film or canvas
awnings.

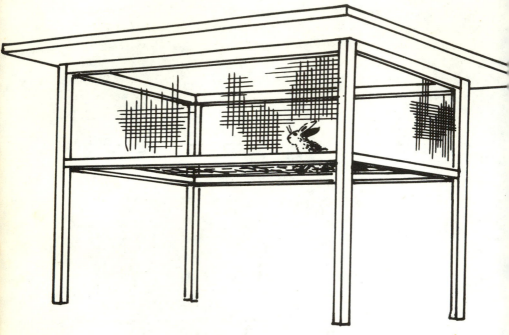

This outdoor hutch has no door because the roof is the door. The hutch is framed of two-by-twos, two-by-threes, or two-by fours and covered with rabbit wire. The roof is a sheet of plywood covered with roofing. It is attached to the frame with two or three butt hinges along one side and secured on the opposite side with two hooks-and-eyes.

Some urban rabbit raisers use a slatted wall-like structure to screen a backyard rabbitry. Others use the kind of decorative privacy fencing seen in many backyards in towns and cities. Still others erect arbors around and over the hutches, growing dense vines over these arbors to shield the hutches from view.

Outdoor hutches *can* be built out of scrap lumber and one-inch poultry mesh, with ⅝-inch hardware cloth used for floors. One still sees these jerry-rigged hutches. But, both show and commercial breeders are backing away from these materials (with the exception of used lumber). Poultry mesh is an extremely fragile wire. It soon rusts out in the corners of cages where rabbits urinate. The ⅝-inch hardware cloth also is not a desirable flooring wire for rabbit cages. The hardware cloth sags rapidly under the weight of a heavy doe and her large litter, and sore feet can result. Moreover, the large droppings of does are often too big to fall through the mesh of the hardware cloth, clogging the wire. Clogged openings in cage floors encourage the eating of droppings and thus increases disease among the young rabbits.

Neither of these wires offers any protection whatever from determined predators. Even a small dog can and will tear its way through either the poultry mesh or the hardware cloth.

Outdoor hutches are usually built of two-by-two-inch framing, with the wire cage fastened to the *inside* of the framing. Any exposed wood parts are going to be gnawed by the rabbits, which exposes the rabbits to germs in urine and feces. It also means cage repairs or replacements at a much earlier date when the rabbits' chewing has weakened the wood to the danger point.

Properly built and maintained, outdoor hutches can last many, many years. Legs for outdoor cages are usually made of two-by-fours, for stability and strength. Roof framing and bracing may be built of either two-by-four or two-by-two-inch lumber. As a rule, doors in outdoor hutches are framed with wood having a metal hook or latch for fastening from the outside.

Wooden parts of outdoor hutches may be left plain, but are much more attractive if they are painted white. The white exterior also reflects the heat of the sun, helping to keep the animals cooler in the summer. The lower twelve to eighteen inches of many hutch legs are painted with roofing tar or creosote, to repel the mites that originate in the soil and love to migrate to rabbits' ears.

One of the major advantages of raising rabbits in outdoor facilities is *free* fresh air circulation. Any odors that may originate under the hutches are diluted and wafted away. Flies are usually less of a nuisance around outdoor rabbitries, so long as reasonable sanitation measures are

If you've got the money to spend, commercially made hutches are available from several manufacturers. These units, made by Favorite Manufacturing Company, have all-wire cages with built-in feeders mounted three-high in all-metal racks. The racks

taken. Of course, no one can expect to allow the manure to pile up indefinitely without having some of the attending odors, even around the fresh-air rabbitry. And the zillions of flies drawn by such odors, as well!

The outdoor hutch, or other rabbit housing for that matter, should be located on well-drained soil if at all possible. Otherwise, the soil soon becomes saturated with urine, creating not only a messy appearance, but a fly breeding ground and health hazards for the rabbits. Urine contains

have catch-pans for droppings and can be equipped with casters for easy moving. For a laboratory or a commercial rabbitry, such units are worthwhile, but for the homestead rabbitry, the expense of them can be out of line with the returns.

nitrogen which, along with bacterial action on manure that stays damp under the cages, is the source of the offensive aroma one sometimes detects around rabbitries. This is known as ammonia gas, and it can cause irreparable damage to the rabbits' respiratory tracts.

There is one disadvantage to using outdoor hutches that discourages some raisers from making use of this method of housing rabbits. The wooden frames of the cages, if not properly installed, are harder to disinfect than all-wire cages. And if allowed to become thoroughly saturated with urine, the wood can become a source of offensive odors. But if the wood-framed hutch is built properly, cleaned often, and disinfected after each cleaning, these problems can be overcome without too much trouble.

Indoor, All-Wire Cages

The all-wire cage can be used in any sort of roofed building: the open pole shed, the pole barn, the garage, or the larger hay and livestock barn. Being made entirely of wire, these cages are easier to clean and disinfect than the outdoor hutch. More of these cages can be utilized in a smaller space than is possible with the wood-framed outdoor hutch.

At one time, wire cages were built with double walls of wire between cage compartments. Today's prices on wire make this impractical in most cases; most breeders using the two-compartment cage now use only one wall of wire between the two compartments. This often leads to fighting between the rabbits in the different compartments, or to fur chewing, which ruins pelts.

For the person wishing to produce more rabbits than he will need for his own use—selling the surplus to others— the all-wire indoor cage is often more convenient for tending the herd during bad weather.

There are other advantages in using all-wire cages, which would appeal to almost any rabbit breeder. They are more easily sanitized, for one thing. A propane hand torch and a brush are usually all that's needed to clean and disinfect the wire. The torch flame is still the surest method of dealing with certain germs which can affect rabbits. It is also the quickest and easiest way of getting rid of hair that accumulates on the wire. Hair is not such a problem with outdoor hutches or on wire cages in an open shed, but it can become a real mess on wire cages in a closed or partially closed building. If it's left on the cages, the rabbits wind up

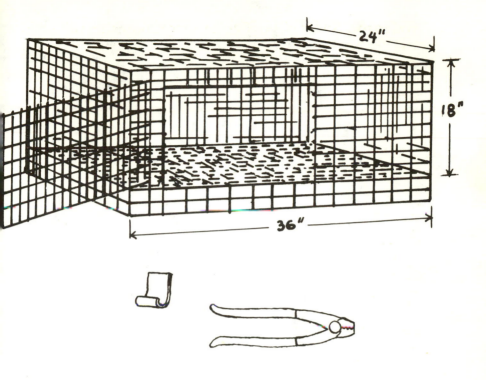

The all-wire cage is widely regarded as the ideal rabbit housing. And the all-wire cage isn't all that difficult to make. The wire is joined at seams using J-clips, which are fastened using special pliers. For each cage, you need eighteen feet of twenty-four-inch-wide wire. Cut off two thirty-six-inch lengths for the floor and roof. Cut off a twenty-four-inch length for the door. Then lay out the remaining piece of wire and, using a board for the straight-edge, bend it into the sides of the cage. Measure thirty-six inches from one end and bend, then measure another twenty-four inches and bend, then thirty-six inches and bend. Use the J-clips to join the ends and attach the roof. Attach the floor six inches from the bottom of the sidewalls, so the cage measures eighteen inches from floor to roof. Cut an eighteen-inch-square hole in one long wall for the doorway, then attach the piece of wire that's the door so that it overlaps the opening on all sides.

eating hair, drinking hair, and breathing hair. Not the best of situations for continued good health in the herd.

Special wires are manufactured for rabbit cages. These are far better, whether the raiser chooses outdoor hutches with wooden frames or the all-wire indoor cages that are hung with strong wires from rafters or ceiling joists in a building. Wire for the sides and tops of the cages is made in a 1-by-2-inch mesh. Floor wire for rabbit cages comes in a 1½-by-1-inch mesh.

These special wires are usually found in a fourteen-gauge or sixteen-gauge weight and can be purchased at most feed companies, from all rabbitry supply houses, and even at those rabbit-meat processing plants that carry supplies for their commercial rabbit producers. Droppings fall through the special floor-wire mesh, making the breeder's job of cage cleaning that much easier, as well as giving some additional health protection to the rabbits.

The outdoor hutch, or the all-wire indoor cage, may be built in almost any size that is comfortable and convenient for the person who will be working with the rabbits most of the time. The cage must provide ample room for the rabbits to get sufficient exercise to stay healthy and active, of course.

Many people prefer the cage that measures thirty-six inches long by thirty inches deep by eighteen inches high. For others, the cage that is shallower—about twenty-four inches in depth—is more convenient. The depth of the cage depends to a great extent on the height and arm length of the person working with the cage. When the depth of the cage is cut to below thirty inches, the length must be increased. A cage measuring twenty-four inches deep by forty-eight inches long provides sufficient room for a doe and a litter of seven or eight young rabbits. The cage measuring thirty inches deep by thirty-six inches long will provide approximately the same number of square feet of space as the shallower, longer cage. A major consideration here is to allow *at least* seven to eight square feet of space for a doe and litter.

The herd bucks can be housed in smaller cages, but they too should be allowed plenty of room for exercise. An idle buck, lying around for lack of room to exercise, becomes a fat buck. And a fat, lazy buck means few if any young rabbits! The buck's cage should be no smaller than twenty-four inches wide by thirty inches deep, or vice versa. These are *minimum* space allowances for a rabbit; larger cages offer more room for activity or for lying apart for better cooling when weather is hot.

The floor of an outdoor hutch should be built well off the ground to prevent rats, mice, snakes, and other similar varmints from gaining easy access. A hutch built about three feet off the ground helps to protect the rabbits from larger predators, too. This three feet of space between the cage floor and the ground prevents all but the largest dogs from damaging the rabbits' feet.

It is usually practical for the rabbit breeder to leave the young rabbits with the doe until they reach butchering weight. Sometimes, though, a litter will fail to reach the desired weight by the time it becomes necessary to wean them so that the doe can have a two-week rest before kindling a new litter. Therefore, it is always a good idea to have at least one spare cage or hutch reserved just for the purpose of finishing out fryers.

For three or four working does, one large hutch should be sufficient to hold any small fryers. A good rule in determining how many extra hutches are needed is to allow one finishing hutch for each three working does. One may also desire to keep one buck-sized hutch available for growing out a new young doe or herd sire.

These spare hutches will be empty part of the time, but they can come in mighty handy when there's a litter to be weaned and held for another week or two. A cage that measures twenty-four-by-forty-eight inches, or thirty-by-thirty-six inches, will comfortably house seven or eight young rabbits to weights of 4½ pounds to 5½ pounds, when they are ready to butcher. Packing in more than this can result in fighting, traumatic injuries, illnesses, and damaged pelts. Even the very young rabbit is a truly vicious fighter when he feels aggression is required to obtain more food, enough space on a hot day, or adequate air circulation on a muggy, humid day.

Buildings

The rabbitry may be located in an already existing structure. A dry, protected section of a barn is suitable for putting up rabbit cages or hutches. A well-built shed is also adequate housing in most areas. Many people, living in urban areas, use garages for rabbit houses, never parking the automobile in the building with the rabbits.

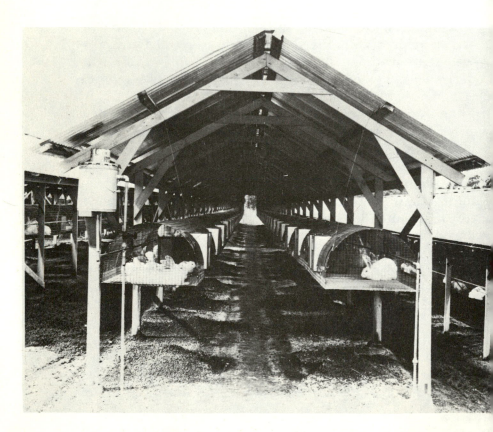

Several points should be taken into consideration in deciding what type of building will be used if it is necessary to erect a special unit for rabbit production. Of major importance is the locality, climate extremes, and the accessibility and desirability of electricity for cooling.

In many regions of the United States, the uninsulated frame building is satisfactory. Where extreme heat or cold must be taken into consideration, insulation will be almost an essential. Thickness of insulation will depend on the severity of heat or cold in the particular area. In our location on the Ozark Mountains, we found an all-wood building to be better than metal.

In our experience with both wood and aluminum barns, we discovered that the wood building was cooler during the summer, even without the use of fans, than the

though the homestead rabbitry isn't going to be as big as these two com-
ercial operations, its building can be designed—on a smaller scale—like the
mmercial ones. In temperate climates, an open-sided structure will shelter
-wire hutches from rain and sun. In colder climates, sides are important to
otect the rabbits from the elements, but ventilation must be provided.

metal barn was *with* the use of fans during hot daytime hours. The rabbits housed in the open shed with a wood-and-composition shingle roof appeared to stay cooler than those in either of the partially enclosed buildings.

A few precautions are necessary if rabbits are housed inside a barn where other, heavier livestock is also kept. Some method of screening the rabbits from other livestock should be utilized, to protect the rabbits from the upset of seeing (to them) some monstrous creature suddenly appear right at their front doors. To the newly kindled doe already keyed up and nervous, even the gentlest horse, cow, or goat sniffing around out of simple curiosity would appear to be a mountainous, malodorous, and fearsome threat to her newborn young.

Rabbits can and will adapt to seeing such large animals when they see them regularly, provided the larger animals *always* act exactly the same, not making any different and unusual sounds or movements. But, while the rabbit doe is adapting to their presence, the breeder can easily lose baby rabbits because of the doe's terrified movement in and out of the nest box.

Consequently, a rabbitry separated from the quarters of larger animals can be worthwhile, even though it may mean a little extra initial expense. The method of separation from the main part of the barn can vary—for example, a partial wall made of scrap lumber would serve to shield rabbits from too-close contact with the bigger livestock.

In erecting any sort of building for rabbits, space should be provided for storage of feeds, nest boxes, cleaning equipment, etc. This extra space may be set up either in the central part of the building or along the front end of the structure.

Cooling

Shade: If at all possible, any structure used for housing rabbits should be located in the shade of trees. Of course, this is not always practical. Many rabbit raisers without benefit of trees shade their rabbit hutches with vines. These are usually trained over trellises or arbors. In using vines to shade rabbits, one should choose plants that are as resistant as possible to insect infestations.

In the southern and central states, the silver lace vine is thought to be a good plant for this type of shading. Some of the ivies are also considered as useful. One should steer clear of the squash vines, due to their insect populations, unless one is prepared to use insecticides. Remember that rabbits, like other animals, are susceptible to poisoning from insecticides.

In some instances wisteria vines are used for shading rabbit hutches, though these are normally a slow-growing plant. Find out from residents in your own area just which vines are hardiest for that particular region and have the least trouble with insects.

Fans: Where shade is not available, use fans or other means of sheltering rabbits at times from the heat of the summer sun. Electric fans are used extensively in many parts of the country. Place the fans so that there is never a draft blowing directly on the rabbits. Drafts at any time of the year can cause respiratory problems in the animals.

In small buildings, one exhaust fan is often sufficient. Place it in an opening in one end of the building, directly opposite another opening in the other end of the structure. This creates fresh air circulation by pulling outside air through the building, rather than blowing air into the barn. The major concern is not so much actually cooling the air as it is keeping fresh air circulating through and around the cages.

In the very large barn, and especially in a metal barn, both intake and exhaust fans may be needed. It is often helpful to close all openings along the sides of the building, using exhaust fans in one end and opening windows wide along the opposite end where the intake fans work. This creates an air flow that benefits rabbits through the full length of the rabbitry.

Raisers in extremely hot areas with low humidity make use of water evaporation and fans. A cooling pad is built into one end of the structure, with one or more fans built into the opposite end. Air is pulled through a wet cooling pad, which lowers the temperature inside the building by several degrees. This type of cooling procedure is not practical in those areas having high humidity most of the time.

Sprinklers: Another cooling method is also useful in certain areas—the sprinkler system. This is a simple procedure if one has a sufficient supply of water. A garden sprinkler is placed atop the building, and when temperatures reach the upper 80s and 90s, water is turned into the hose. The pressure is kept fairly low so that water tends to dribble down over the roof, dripping off the eaves. This will offer some measure of cooling to the building.

Awnings: One of the simplest and least expensive cooling methods, we believe, is the awning. We made our own rabbitry awning—out of white cotton feed sacks. The emptied sacks were ripped open and washed. They were then sewn end to end with twine to make the desired length, in our case, 25 feet.

On one long side of the resulting strip of material, we made a two-inch hem, into which we inserted a damaged one-inch plastic garden hose to act as an anchor. The ends of the garden hose were fastened to the corners of the building with heavy wire. This kept the awning from flapping against the poultry mesh covering the open upper half of the wall, which would have frightened the rabbits in the row of cages nearest the wall.

This homemade awning was then tacked to the ends of the rafters with large-headed Sheetrock® nails. Such an awning can also be made of burlap bags. An awning can be used in very hot, dry areas as well as humid ones, and in conjunction with a roof sprinkler system.

Winter Protection

While domestic rabbits can tolerate quite a lot of cold weather, they do not tolerate *cold drafts* well at all. Before winter winds start blowing, check your building (or outdoor hutches) thoroughly to see if there are any spaces through which drafts might blow.

If there is a crack—even a small one—in a wall near a cage, it should be covered. In some instances, a narrow strip of wood will do the trick. If a really bad break is discovered in a board, it would be wise to replace the board.

The open-sided building presents larger problems, but in many areas the open building is more practical due to longer summers. If the building is small, the four-mil plastic sheeting that is often used to cover dwelling windows can be utilized to protect rabbits from cold drafts during winter months.

For larger buildings, it is sometimes more practical to buy the opaque plastic curtains that are used in large poultry buildings. These can be bought in any length (since the material is put up in bolts), with cord to anchor the curtains against the windows. A small winch and cable system is usually used to raise and lower the curtains, but in the smaller building (less than forty feet in length), the curtains can be raised and lowered manually.

A metal pipe is run through a hem at one side of these heavy curtains to make them more stable and easily handled. Unlike the clear, four-mil plastic sheeting, these opaque curtains do cut out some of the natural light of the sun.

It is fairly simple to offer rabbits in outdoor hutches protection from cold wind and weather. Many people in colder climates make use of the hinged wall, a wooden wall attached on the outside of the wire-cage wall. These hinged walls are attached in such a way that they can be raised or lowered when temperatures warm up. The hinged wall that can be raised offers an additional shaded area around the hutch when opened during warm weather.

In areas where it is possible to use the open, pole-type shed, covering the exterior of the shed with sheet plastic is usually all that is required to keep the rabbits comfortable. The open shed can also be protected by portable walls that are removed during summertime. With rabbits, the most important thing is not so much keeping the animals *warm* as it is protecting them from *drafts of cold air*. Protected from drafts, the adult rabbits' fur insulates their bodies. Smaller rabbits should be allowed to keep their nest-box bed until they are about five to six weeks old during very cold weather, to give them additional protection against cold.

In extremely frigid areas, where temperatures plunge to -40 to -50 degrees below zero Fahrenheit, a rabbit building should be well insulated, of course, if breeding is to be continued through the winter. Outdoor hutches can still be useful in these areas but must be specially constructed to fend off the cold. Raisers in very cold areas can benefit from using a "double wall" hutch construction, using insulation as needed between the two parts of the hutch walls. Placing hutches on the south side of some larger structure, such as a barn, or setting hutches on the south side of a hedge or tree windbreak also helps to keep the rabbits comfortable.

Equipment

Feeders: There are several types of feeders which are acceptable for use in the rabbitry. The most popular type with raisers is the all-metal, filled-from-the-outside feed hopper. This may be purchased from rabbitry supply houses. It can also be handmade at home, if one is handy with metalworking tools such as tin snips, metal shears, rivets, or soldering irons.

Some of the feeders described herein are not at all practical in larger rabbitries. For the small homestead raiser however, handmade feeders are both useful and less expensive than manufactured ones.

One such homemade feeder can be manufactured by

cutting down one side of a one-pound coffee can to about two inches from the bottom, then halfway around the body of the can. This leaves about two inches of metal intact near the bottom of the can. Fold the cut part (it will be a sort of three-cornered portion of the side) flat across the opening. Crimp the sharp edge of the cut side back around the body of the coffee can. The edge may be held in place with small rivets, or a piece of strong, smooth wire can be pulled tight and tied into place in a pinch. Soldering along the edge of the folded-over metal would also work well.

With the now-flat side of the can facing you, bend about ⅛- to ¼-inch of the sharp edge of the lower portion (the feed trough) of the can inward, and crimp it down snugly with a pliers. This is to protect the rabbits' mouths from being cut by the sharp edge of the metal.

To fasten the feeder to the side of the cage, make hooks of the same general type that are used on the factory-made metal feeders. Just punch a small hole on either side of the can directly behind the flat portion and about one-third of the way down from the top of the can.

Then, insert a heavy wire through the holes across the breadth of the can. Clothesline wire works well for these feeder hooks. Bend the ends of the wire forward at the sides of the can. Insert the feed-trough portion of the can into the opening cut in the side of the cage, and bend the ends of the wire at the proper length to form a hook to fasten snugly over the cage wire.

The coffee-can feeder may be left plain or painted, depending on the raiser's preference. If the coffee can is painted to give feeders a neater, more uniform appearance, be absolutely sure you are using a *nontoxic* paint, and preferably one that is fast drying.

A hanging feeder can be made by using a coffee can with only the ends cut out, and an ordinary round or octagonal, aluminum cake pan. This is very useful in the cage where a doe has an extra-large litter, or in a finishing cage where too-small fryers are being fattened out; it lets numerous fryers eat at the same time.

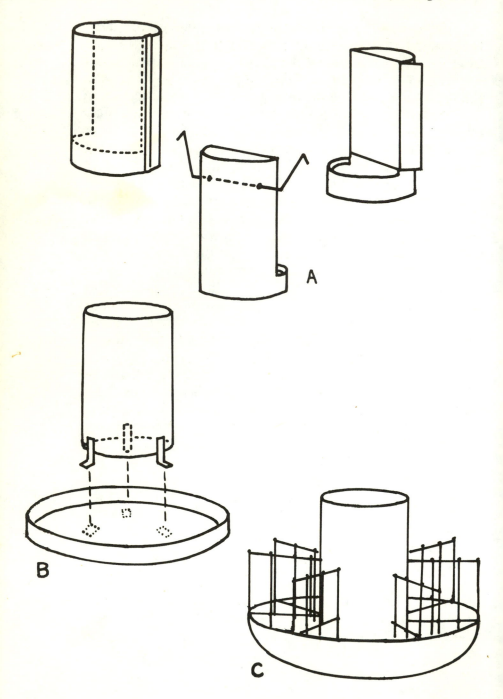

A

B

C

The coffee can is riveted to three thin strips of metal; these metal strips are in turn riveted to the bottom of the cake pan. The two utensils should be separated by about one inch of clear space, so that feed can slide down into the cake pan, which serves as the feed trough.

After the can and the cake pan are joined with rivets, a small hole is punched in each side of the top of the can. Strong wire is fastened through the holes, forming a bail such as those on buckets. Another strong wire is attached to the top of the rabbit cage. Loop the lower end of this wire around the "bail" of the feeder, being sure that the feeder can be reached easily by the young rabbits but not be low enough to allow them to climb into the feeder trough and soil the feed.

Some feeders are handmade of part wood, part metal. When wood is used in making feeders, no wooden parts should be accessible to the rabbits. It doesn't take a litter of rabbits long to chew through the sides of an all-wood feeder. The trough, or hopper, of a part-wood feeder should always be made of smooth metal. A wooden feeder is not only more trouble to make; it is also less sanitary than the all-metal feeder.

Coffee cans can be turned into a variety of rabbit feeders by the handy homesteader. As shown in A, the can can be cut along the seam and around the circumference near the base. The resulting flap is bent flat across the new opening and pop-riveted to the side of the can. A length of wire run through two holes in the can is used to secure it to the hutch. Figure B shows how to make a freestanding feeder. Cut out the top and bottom of the can, pop-rivet three right-angle tabs to the sides, then to a round cake pan. Be sure to leave space between the bottom rim of the can and the flat surface of the cake pan. In figure C, the can is fastened, as in B, to the bottom of a metal bowl. Then pieces of wire fencing are soldered to the can and the bowl to split the bowl into feeding sections.

If one is really handy with tools, a very professional job can be done in building all-metal feeders. A good metal cutter, a soldering iron, rods of solder meant for use with galvanized metal, a piece of heavy wire (for feeder hooks) about eight inches longer than the feeder will be, and a pair of pliers are all that is needed.

Metal for handmade feeders can be bought new, or it may be salvaged at building sites. Builders frequently discard many large scraps of perfectly good galvanized metal or sheet aluminum. Some conservation-minded people salvage these scraps and make very nice rabbit feeders of them.

Crockery feeders are often used, and they serve a good purpose when used in the right place. For a single adult rabbit, a small crock makes a good feeder. In a cage with many young rabbits, using a crock as a feeder is asking for trouble with diseases. The young rabbits find that the crock makes an excellent place to sit to keep their feet warmer in the winter. At least part of the litter is going to decide that the crock also makes a dandy bathroom! The resulting soiled and wet feed is an ideal breeding ground for germs, and the little rabbits will invariably go ahead and eat some of the contaminated feed.

Waterers: Rabbits require water, and lots of it, to do their best jobs of producing, milking, growing, and staying well. Water must be provided for the full twenty-four hours in the day, year-round. Freshness of the drinking water is of utmost importance. There are several acceptable ways of providing fresh water for rabbits at all times.

Most popular among commercial and show breeders these days are the so-called automatic watering systems. Having used one of these systems for several years, I would caution anyone contemplating them that these methods of watering rabbits *are not by any means completely automatic*. The self-watering systems must be cleaned frequently, if they are to remain among the most trouble-free watering methods.

Cleaning of the automatic system should be done at least once a week, and twice-a-week cleaning is more de-

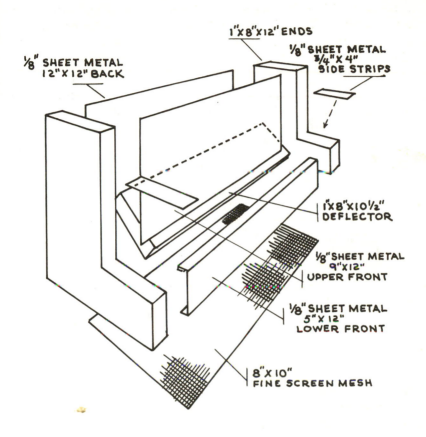

1"x8"x12" ENDS

⅛" SHEET METAL
¾"x 4"
SIDE STRIPS

⅛" SHEET METAL
12"x12" BACK

1"x8"x10½"
DEFLECTOR

⅛" SHEET METAL
9"x12"
UPPER FRONT

⅛" SHEET METAL
5"x12"
LOWER FRONT

8"x10"
FINE SCREEN MESH

A high-capacity feeder much like the commercially made models can be fabricated from wood and metal as shown. Cut the parts and fasten them together. The trough is slipped through a rectangular hole cut in the side of the hutch so the rabbit can feed easily. The hopper remains on the outside of the hutch, easily accessible to the homesteader. Any fine matter in the feed will sift through the screened bottom.

Commercially made hayracks are available, but it is
easy to make one. Construct a small box of ¼-inch
plywood, leaving the top open. Attach a small tab
the center of the bottom of the opening and run a
length of wire through a hole in end at the top. Rest
the tab on the hutch wire, swing the opening of the
hayrack against the hutch and hook it in place with
the wire.

sirable in maintaining as disease-free an environment as possible. One should also know that research crews working in rabbitries in California and in the Ozark Mountains area have proven that *contrary to prior beliefs, bacteria can swim against the minimal current pressure* which must be used in the pipelines feeding the automatic system. Disease germs can travel via the watering system, from a dewdrop valve furnishing water to a sick rabbit to the dewdrop valve in another cage housing healthy rabbits.

The water crock is not just a poor solution to keeping water before the rabbits at all times; it is probably one of the

All-metal feeders such as this can be purchased from a number of manufacturers. Or, if you are handy with metal, they can be constructed in the homestead shop. In either case, the feeder in actual use is attached to the hutch in such a way that the trough juts into the rabbit's domain and the feed hopper is outside the pen.

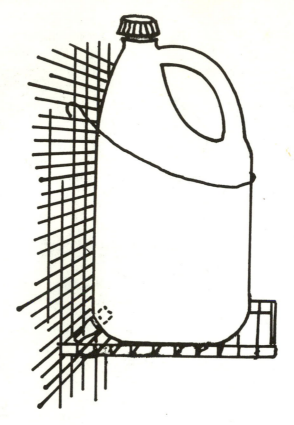

A waterer made from a plastic jug and a dewdrop valve.

most common culprits in the spread of disease. Insects have ready access to the water in a crock; rabbits defecate into the crock; dust, hair, and dirt in the air settle on top of the water.

One of the best individual waterers I've seen used in a small rabbitry was made of a jug fitted with a dewdrop valve. Using a simple plastic bleach or vinegar jug, put a small hole in the heaviest part (called the "bead") of the jug near the bottom. The hole should be placed so that when the dewdrop is inserted and tightened down, the valve protrudes into the cage at an angle.

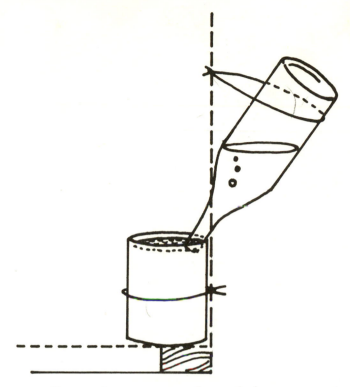

*A really simple waterer can be made from a peanut
can and a soda bottle. Wire the can to the hutch on
the inside. From the outside, insert the neck of the
bottle through the hutch-wire into the can. Wire it to
the hutch.*

The plastic jug is set into wire hoops fastened to the
side of the cage. Or, it may be set on a wire shelf which is attached to the side of the cage. The upper portion of the jug
may have to be anchored by passing a wire around the jug
and through the cage wire, so that the rabbits can't push a
partially empty waterer out of reach, denying them the
water they must have. By using the plastic jug waterer, one
can medicate water for one litter, or for one single rabbit. In
using the automatic watering system, one must either cut
off access to a dewdrop and give a sick rabbit a separate
crock of medicated water, or medicate the entire herd,
which isn't always a wise thing to do.

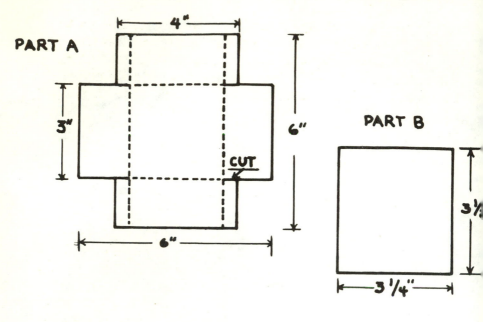

PART A

4"

3"

6"

CUT

6"

PART B

3½"

3 ¼"

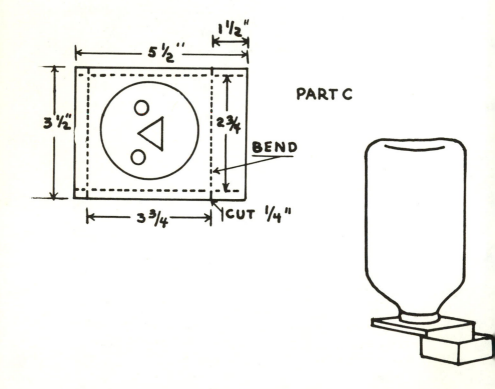

1½"

5½"

PART C

3½"

2¾"

BEND

3¾

CUT ¼"

In warm climates, a satisfactory waterer can be made out of a one-gallon glass juice jug. This type of watering system calls for some talent, once again, with metal cutting and soldering. The glass jug waterer is not at all practical however in cold climates where the water would freeze. Faced instead with a recycled plastic waterer full of ice, one can break up the ice by bumping the jug against a wall or post; the ice then melts much more rapidly than if it remains solid. The plastic jug won't break when the water inside freezes and expands, as a glass one will do; the plastic jug will bulge outward.

Nest Boxes: A nest box is, contrary to the beliefs of some, one of the most important pieces of equipment in the rabbitry. Does have been known to try to build a nest on a cage floor, and to kindle (deliver) their young on the wire, rather than use the half-baked nest boxes they were given.

There are several types of nest boxes, and most of these have considerable merit depending on the area and climate in which they are used. For the very cold climate, one should never try to use the all-wire nest that has gained popularity in commercial rabbitries in recent years. The all-

An intricate project for the metalworker is one that will turn a glass juice bottle into a rabbit waterer. Form a drinking cup from a piece of galvanized metal by cutting and bending as shown (part A). Solder the seams carefully. The "cup" should be 3 inches square and 1½ inches deep. Part B is the bottom of the water-feeding unit. The top of the water-feeding unit is part C. Make four ¼-inch cuts and bend along the dotted lines, lapping the tabs and soldering all the seams carefully. Pop-rivet the lid of the bottle to the top of the unit, as shown, then punch a hole (the triangle) through the lid and unit top. Solder the bottom (B) in place. Slide the apron down into the drinking cup (it should not reach the bottom) and solder the two parts together.

wire box does a fine job in warm areas, and even in temperate zones when it is used inside a well-insulated building. In the average rabbitry, with little or no insulation, baby rabbits will be frozen to death in the wire nest box.

Metal nest boxes with removable Masonite® bottoms are useful in some areas, especially those places having very low humidity. In regions with high percentages of humidity, metal boxes may sweat inside, causing moisture-related disease problems in the very young nestlings.

The wooden nest box is probably one of the most widely used types. For warm to moderate climates, this is usually built of ¼- to ½-inch plywood. In very hot areas, many raisers use window screening to make the bottom of the wooden box, which gives excellent air circulation for the baby rabbits on an extra hot day. The screen bottom also helps to keep the doe cooler when she goes into the box to kindle her young or to feed them later on.

In moderate climates, the wooden box usually has a wood bottom. As a rule, the box in these areas has no top or sitting board across the top; the top is left open to allow more air circulation. Three or four holes, measuring about ¼-inch in diameter, are drilled in the bottom of the box near the back (the highest end), providing drainage to help keep the nest neat and dry.

In very cold climates, many breeders use one-inch lumber for building nest boxes and add a sitting board across the back of the box, which provides a partial cover for the nest. The partial cover extends from four to eight inches over the top of the box, giving ample room for the doe to bed her young underneath. The sitting board should be taken literally, as does love to rest there, and little rabbits out of the nest find it a marvelous place to sit on, play on, and go to the bathroom on! Tops of covered nest boxes must be cleaned off often or they can become filthy in short order.

For a medium-sized rabbit doe (New Zealand, for example), the ideal nest box measurements are: length, about twenty inches; width, about twelve inches; height at

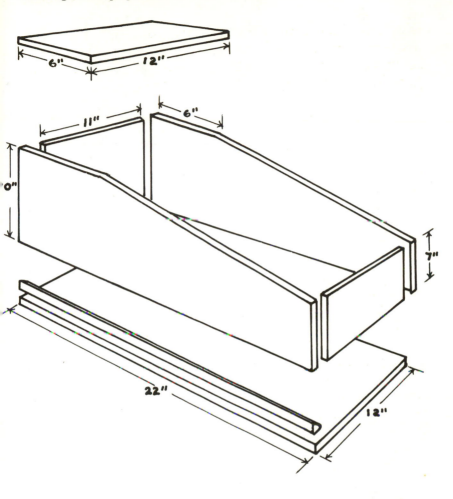

A nest box is a vital piece of equipment, and one that is easily made from a piece of plywood. The parts are cut to the dimensions indicated and nailed together using 8d nails.

front, seven inches; at back, ten inches; at sides, ten inches near the back end of the box and sloped down to the seven inches at the front end.

If one of the Giant breeds is worked (Flemish Giants, for example), the nest box should measure around twenty-two to twenty-four inches in length, with a width of about twelve inches as for the medium breeds. Box height can be the same for the Giants as for the medium-sized rabbits. If one works one of the Dwarf or near-Dwarf breeds (such as the Netherland Dwarf or the Dutch), the nest box should be scaled down accordingly. The tiny does can get into a larger box, but it isn't convenient as if they had a box built for their own size.

Handy Small Tools: Some people go overboard in buying the tools and equipment they think they'll need in producing rabbits, and wind up with more money invested than their rabbits can possibly pay back in a reasonable length of time (unless the rabbitry is operated on a strictly commercial basis with large sales of breeding stock or laboratory stock). The job can be well done in the small herd without large cash layouts for ultramodern tools.

Take feed scoops, for one example. Most of the metal scoops I've seen in livestock feed and equipment stores sold for close to five dollars. Less handsome, but just as efficient when it comes to scooping feed into a rabbit feeder, is a homemade scoop fashioned from another of those plastic jugs.

Starting below the handle of the jug with a sharp knife, cut out a diagonal section of the side of the jug. Angle the cut toward the front of the jug (opposite the handle), then cut around the bottom of the jug just above the heavy bead or ridge. A one-gallon jug will make a scoop just right for filling the large doe-and-litter feeder with one load. Using a smaller jug—perhaps one quart or even smaller if you can find one—results in a scoop that will make filling smaller feeders easier.

For those bucks and single does without young, an ordinary, metal kitchen measuring cup works extremely well.

One precaution: Be sure to weigh the cup empty, then weigh it again with feed in it, to assure that your rabbits are getting the required amount of feed per day. Measuring cups are usually meant for measuring liquids. Pelleted feeds and liquids will weigh different amounts.

There come those days in every rabbitry when there simply isn't time enough to remove all the droppings from under cages with a shovel. To help keep the rabbitry floor looking as neat and clean as possible, make use of the old leaf rake. These lightweight rakes are ideal for sweeping up spilled straw and nesting material.

On those heavy cleaning days when the manure must be moved out, instead of breaking the back using one of those woman-killer manure forks or shovels, try the little potato fork from the garden tools. It won't move such an enormous load at one time as the huge manure fork, but it does the job almost as fast, and is a lot easier on the back muscles.

For those small, day-to-day cleaning chores, the common bathroom toilet-bowl brush is surprisingly useful. It is inexpensive compared to the steel wire brush and, to me, much easier to use. It is excellent for knocking loose droppings that hang on the wire floors with hair. Just whisk the brush across under the wire of the floor and hair-dung droppings pop loose.

The small propane torch is one of the best timesavers a rabbit raiser can have. For the small rabbitry the hand-held torch is ample. When a cage needs a complete cleaning and disinfecting, the torch flame is one of the most effective methods.

Housing Sanitation

Sanitation of hutches and buildings is one of the most important factors of rabbit husbandry. Experienced rabbit breeders seem to harp so much on this subject. They know that caged-up rabbits require special care and cleanliness to prevent sometimes-disastrous outbreaks of disease. Those items one reads in magazines about care and sanitation really are not so many old wives' tales! Confined in cages, rabbits are more susceptible to disease caused by filth and neglect than if they ran loose like their wild cousins.

Hutches should be thoroughly cleaned and disinfected regularly. Young rabbits kept in a dirty cage are almost assuredly going to develop coccidiosis, one of the most damaging rabbit diseases. The doe housed in a dirty cage is more likely to suffer stresses than the doe in a clean cage. A buck housed in filthy quarters can pass disease germs along to does he breeds.

Keeping cages and buildings clean isn't really as hard as it may sound. If cage floors (the likeliest place for soil to accumulate) are brushed clean every few days, the regular thorough cleaning is made easily and quickly.

For the complete cleaning of a cage, the propane torch is considered by experienced people to be the most reliable means of killing all disease germs on cage wire and on wooden parts of outdoor hutches. It is the only method presently known that is a sure kill of cocci germs. Where washing of cages with a disinfectant in water mats the loose hair on the wire and makes it next to impossible to remove, the propane hand torch quickly and efficiently singes the hair off. A brush rapidly removes any ash left by the burned hair and lint.

For the in-between minor cleanups of cage floors, a quick trip through the rabbitry a couple of times a week with a brush works wonders on lessening the hard work of a complete cleanup. Just whisk the brush across the underside of the wire floor to knock loose any droppings that may be caught on the wire with hair.

Each spring and fall, cage floors should be checked for the beginnings of rust spots. This occurs most often in a corner of the cage where the rabbits urinate. If rust spots are beginning to show, the damage can be slowed considerably by doing a quickie paint job with fast-drying aluminum paint. The whole floor may be painted if desired, but a quick touch-up in only the affected area can save a floor replacement in many cases.

Anytime the propane torch must be used for a complete cleanup on a cage, the rabbits should be removed to another cage. Giving the entire cage a thorough going-over, inside and outside, is the best insurance you can get for preventing the spread of diseases.

Sometimes it is quicker and easier to clean a really filthy cage where fryers have scoured for some reason, with a liquid cleaner before using the torch. Lysol® is a good cleaner for this purpose. Pine-Sol® is also good, but in our experience, some rabbits dislike the odor of Pine-Sol®. None ever showed any aversion to the scent of Lysol®.

Many rabbits, especially does, detest the odor of chlorine bleach being used around their cages. Ammonia is also not a good cleaner to be used around the rabbitry. If lye or any sort of soap or detergent is used on rabbit cages, they should be given a thorough rinsing in clear warm water to get rid of any residue. Small rabbits will lick and chew on the wire. Lye could eat into the tender parts of the mouth, and some detergents will cause illness in anything swallowing them.

The nest box is one place where torching is not considered adequate for proper sanitation. To clean a box properly, first remove the bedding. Scrap out any soiling and damp bedding from the bottom of the box. Hose thoroughly, using one of those high-pressure garden spraying attachments, if possible. When weather is agreeable, drying in sunlight and fresh air is very beneficial in removing odors and killing any lingering germs.

When the box is thoroughly dry, set it aside in the rabbitry where it will remain clean and dry. Just prior to stuffing the box with fresh straw or other nesting material for the next doe, run the propane torch over the box (inside and outside), paying special attention to seams in wooden boxes. Then give the box a light spraying with Lysol®. If you are adverse to using sprays of any kind, dip the box in a solution of Lysol® and water after it is hosed out, then dry it in the air and sun. The Lysol® gives a little added protection to those vulnerable baby rabbits when they come. We always preferred scraping, hosing out, and drying the boxes upside down on a rack for several days out-of-doors. Then before stuffing the boxes with straw, we sprayed them with Lysol®—all over, inside and out—until the wood (plywood) was just slightly damp.

After the Lysol® dried, we stuffed the boxes and immediately delivered them to due-to-kindle does. Some breeders prepare them in advance, especially in very large herds, but we figured that the longer the boxes and nesting material are allowed to set, the more chances that mice, insects, and germs could invade the boxes.

Crocks used for feeding and watering rabbits should be given a thorough cleaning and disinfecting at least once a week. During hot summer weather, cleaning them more often helps insure better health for the animals. One of the best methods of cleaning small equipment like this is to wash them in a good detergent and water, then dip them into a solution of disinfectant and water. Let the feeders dry before being used again, since any residual moisture will spoil food. Water crocks need not be completely dried, but should air long enough that the odor of the disinfectant dissipates somewhat.

If one uses an automatic watering system, this should be drained at least once a week year-round, then flushed out with chlorine bleach or a special sanitizing material called Sanitizol® (available from rabbitry and poultry supply houses). This draining and flushing removes any residual contaminants left by minerals or anything else in the water. It also helps to minimize the dangers of disease germs introduced into the system by ailing rabbits.

Why is this special attention needed with a watering system that is supposed to be automatic? In certain parts of the country water leaves a residue in pipes, even when the pipes carry the high water pressure used in households. In the rabbitry watering system, the pipelines only carry about two pounds of water pressure. In effect, this means the water in the pipe is hardly moving at all, resulting in almost stagnant water. During summer months bacteria grow rapidly in slow-moving water, creating a need for even more astute care of watering equipment.

Here in the Ozarks area a few years ago, rabbitry operators were surprised to find that their automatic watering systems which hadn't been cleaned regularly contained a clear, slimy, jellylike substance. This is considered very unhealthy for any kind of animal, causing diarrhea. Weekly flushing of the lines prevents accumulation of this material. While this particular residue may be a phenomenon in the Ozarks region alone, other areas may have their own problems with various minerals, for example.

A tub or vat of some sort is a might handy place in which to wash those crocks and feeders each week or more often as needed. We found one of the old-fashioned laundry rinse tubs to be an excellent help with these chores. These rinse tubs usually have legs just about the right height for most people to work with them, and most of them also have built-in drains which make disposing of soiled water a snap.

If there is no laundry rinse tub available, a regular washtub may be used. A bench on which to set the tub or tubs should be built at whatever height is most convenient to the person doing most of the equipment washing. Unless fitted with some sort of drain hose, this washtub can be mean to empty. If your tub is located inside the rabbitry, the soiled water from either type should be channeled by way of drain hose or pipe to the outside of the building, or carried out. Excess moisture inside the rabbitry often results in various kinds of diseases for the rabbits, besides being messy to tramp around in while doing the chores.

Housing sanitation simply means that old job of every animal raiser: manure shoveling and hauling. In a rabbitry, it is very important to keep the manure cleaned out from under cages, whether indoor or outdoor. In fact, this is one of the most important parts of raising healthy rabbits. If allowed to pile up, the manure becomes a breeding ground for flies. Places that are continually wet with urine are ideal blowfly hatcheries. Another, just as important, factor of manure accumulation under cages is the ammonia gas the combination of urine and manure creates. Ammonia in a rabbitry can be the seat of more problems than any other one thing.

When a building is used to house rabbits, one soon discovers that every opening that is covered with screening or poultry mesh becomes coated with hair, dirt, and lint. Here again the small propane torch is a lifesaver. A quick burning with the torch removes the hair and lint, opening up essential vents for air circulation. The resulting ash and dust can then be removed with a brush, leaving a neat, light, and airy-looking rabbit house. Left to collect on these openings, the hair and lint eventually would completely choke off circulation of fresh air into the building.

Chapter 2

Foundation Stock

The New Zealand rabbit, left, and the Californian, right, are among the meat-type rabbits that the homesteader will be looking for to stock his rabbitry.

Choosing Breeding Stock

There is a long-standing argument about just what kind of stock a new breeder should obtain to form the foundation of his rabbit herd. Those people who are mainly interested in putting their best rabbits into competition on the show tables naturally say that every beginner should start with the "best" rabbits. The best being top-notch, expensive show

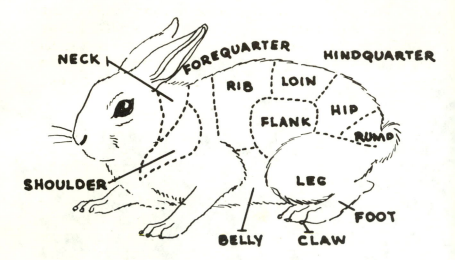

The different sections of a rabbit.

animals, of course. Yet those men and women who get right down to the nitty-gritty of making a living in rabbit production say that they "wouldn't have a show rabbit in their herds."

The type of stock the new raiser buys depends on just what he wants to do with his herd. If he desires to show rabbits, then he should invest in the best show stock he can afford, by all means. The person wanting nothing more than to put meat on the table should buy stock from the best production herd he can find.

Registered show rabbits may have a 1 or 2 percent better dressout ration than the strictly commercial production type of rabbit. But, the buyer is going to pay dearly for that registration number in the right ear and the piece of paper that proclaims the rabbit has been registered with the American Rabbit Breeders' Association.

Rabbits are not judged in the same manner as goats, dairy cattle, or swine. Conformation, fur, and condition are the points that count in rabbits. The judge never knows whether the "best" doe in the show has produced numerous rabbits or no rabbits. It doesn't really matter whether the doe is a good milker and mother; only her conformation, fur, and condition are judged.

Therefore, I believe that for the rank beginner, buying expensive show stock to start a home-meat rabbitry is unnecessary. If, after working with production rabbits for a little while, the beginner would like to go into fancy rabbits, he can always add a few show-type rabbits to his herd.

But, he shouldn't really expect to feed a family with the young produced by show rabbits. For one thing, a doe that is kept in top show condition for a good part of each year cannot be worked *and* kept in the best possible shape for showing. Hair made ragged by fur pulling for nest making is enough to put the doe off the show table in a hurry.

There are, of course, strictly commercial rabbit shows, but these are less numerous than the exhibition held exclusively for the show-type rabbit. The strictly commercial show, where one may present a doe with a beautiful litter from tiny nestling to near-weaning age, is most often found at small county fairs. In these small fair shows, it is not necessary to have the doe butterball fat as is the case in the strictly show-stock exhibition. The rabbits can't be snake-skinny of course, but it isn't expected that a nursing doe be fat as can be.

Any domestic rabbit has edible meat, regardless of size, color, or breed. Sizes vary from those called the dwarfs to the huge rabbits called the giant breeds. Poundage varies from about 3½ pounds for the Dwarfs, to around 20 pounds for the Giants. Colors run from the albino with ruby-colored eyes, to the jet black with dark brown eyes.

The most commonly used size for meat production is the medium breed. This size encompasses numerous breeds of rabbits: the New Zealand (red, black, and white), the Californian, which has white body fur with black or brownish tail, feet, and ears, and the Palomino with golden or fawn-colored fur and dark eyes. There are also the Martens, Silver Fox, and numerous others of many different colors and only slightly varying weights at maturity.

The larger the bone structure of the rabbit, the smaller the percentage of meat to waste. Thus, the medium-sized breeds are usually preferred in meat herds to the giant breeds.

The mature New Zealand rabbit weighs more than you think. A New Zealand doe will normally weigh, at full maturity and in good condition, from 10½ to 12 pounds. Now and then, one may see a doe that weighs more than 12 pounds, but this is not common for the New Zealand. The New Zealand buck, at full maturity and in good working condition, will weigh from 9 to 10½ pounds. As with the does, a buck may occasionally be heavier than normal without being over-fat. And, sometimes a buck weighs less than 9 pounds, yet is a really fine herd sire. The doe and buck both tend to throw their size to offspring, thus most raisers try to save stock from their largest animals, so long as those bigger rabbits also throw stamina, productivity and good health to their young.

Unless one intends to show his rabbits, the weight is not so important to the homesteader, as long as the animal is reasonably close to the normal accepted minimum. Over-weight or a lack of accepted size and weight can disqualify a rabbit on the show table, but make no difference in the quality of the meat the rabbit produces.

In fact, I have worked smallish New Zealand does that were far superior to their heavier sisters in nest building, fryer finishing, milk production, stamina, and the ability to work year-round. I've seen does that never weighed more than ten pounds, except when they were pregnant, kindle and raise out well over 200 pounds of fryers in one year. This was, of course, on one of the stepped-up commercial breeding schedules, but the does were in excellent condition due to proper feeding and care.

The homesteader concerned with putting meat on the table shouldn't turn his nose up at a doe because she looks small. She may very well turn out to be the best producer and milker in the entire herd.

Often, the large, very heavy doe has trouble raising out all of the babies she kindles. Being heavier, it is harder for her to get in and out of the nest box at feeding time without stepping on and injuring one or more of the babies. The lighter weight doe rarely damages a baby rabbit by accident in entering or leaving the nest. This doesn't mean that she never accidentally steps on a baby rabbit. It simply means that she is lightweight enough that her foot doesn't crush the life from the tiny rabbit.

I've never known of a dwarf-breed doe accidentally killing one of her young while getting in or out of the nest to feed them. These breeds, the Black, Chocolate, Blue Dutch, or the Netherland Dwarf, rarely weigh more than five pounds at full maturity. They are not normally kept in the strictly meat herd, and never in the commercial meat herd, due to their small size. But their meat is just as edible as that of the larger breeds. Their pelts are as usable as the big pelts; it merely takes a greater number of them to make a given item than if the large pelts are used.

Pre-Purchase Inspection

When buying stock for meat production, ask to see production records of the parents of animals offered for sale. If the selling breeder refuses to show any records, go find stock elsewhere! Any breeder who has a really good herd is always happy to show off his breeding and production charts.

In choosing which young rabbits to start a herd, thoroughly look over the rabbits offered. Start with the fur. The fur on even a young rabbit should not be soft and downy. Run the hand lightly up the back from the tail to the neck. If the fur snaps back into place, the rabbit has good fur. Should the fur stay ruffled up and messy looking, don't buy. The pelt will look just as messy when cured for use in clothing trim, toys, and other items.

Check the rabbit's teeth. The four front teeth, two uppers and two lowers, should be straight and strong. The two upper teeth should overlap the lower ones. Steer clear of a rabbit that has lower teeth lapping over the uppers; this is called malocclusion or buck teeth. The rabbit having such teeth can't eat properly, and is almost sure to throw the same defect to its offspring when it reaches breeding age.

Look at the rabbit's eyes. They should be clear and bright, whether the eyes are the pink of albinos or the darker shade of the colored breeds. If there is an opaque appearance to the eyes, chances are the rabbit is blind, or partially so. If the fur just below the eyes is wet and matted, leave the rabbit alone; it probably has a problem called conjunctivitis, which is a contagious form of cold congestion.

Turn the young rabbit upside down, or ask the raiser to do so, and check the fur on the bottoms of the back feet. On a well-padded rabbit, the fur on the soles of the back feet should be very dense and slightly coarser than the body fur. When the pads on the back feet are thin and soft, the rabbit is quite likely to develop sores on the feet called sore hocks, and will be at best only a fair producer. On a buck, sore feet interfere with breeding. A doe having sore hocks may be unwilling to breed; if she accepts service and conceives, she can have trouble getting in and out of the nest to care for her young.

Conformation (body shape and balance) is important in the meat herd, but not really as singular a requirement as in show stock. To choose a breeding buck (also called a working buck), look for a good chunky body type. The head should be wide and somewhat short in appearance. Male rabbits kept for breeding purposes should have a meaty appearance overall. Shoulders and hips are to be wide and deep. The hips should slope smoothly down to the tail instead of chopping off. Ears should be fairly short, very straight and strong, and usually somewhat thicker in appearance than the ears of a good doe.

In choosing a production doe, look for a slightly longer body type than in the buck. Ears, of course, should be straight and held up properly. With the exception of the Lop breed of rabbits, pendulant ears are considered a genetic weakness which will be passed to offspring.

A production doe should have a full, ample rump on her. Thin hips (sometimes called pinched) can mean difficulties in kindling. The head of a good doe is often a bit longer and narrower than the buck's head. Ears may also look a little longer than the buck's ears.

One method of choosing does that have the least amount of trouble in kindling when they mature which we always found helpful was checking the width and depth of rib cages. When I selected a young breeder from litters raised, I made a practice of placing my hand, spread to full finger span, over the rib cage. Then the hand spread in the same manner was placed over the hips at the fullest point. If the ribs seemed narrow compared to the hips, the rabbit was placed in the meat pens.

This procedure was repeated as the rabbits grew, and any in which the rib cage seemed to be considerably more narrow than the hip were removed from the young breeder cages. By the time a young rabbit was three months old, I expected its rib cage to almost fill the spread of my hand from thumb to tip of the middle finger. A fully mature doe's rib cage was expected to fill the hand span. (I have a seven-inch hand span, which is probably about average for a woman.)

When choosing very young stock to grow out as breeders, the legs should be checked for strength and straightness. Hold the rabbit by the loose hide and fur on its shoulders and neck, then turn it upside down so the feet are easily seen. The back feet should track straight with the body median. If the hind feet stick out to the sides, the rabbit is more prone to develop sore spots on the hocks or callouses from uneven pressure. The front legs should have no bow at the knees. This trait is attributed to genetically weak bone structure and is usually passed on to offspring.

There are various age preferences in buying young stock. Some people prefer buying very young animals, at about ten weeks of age, and raising them out to breeding age. Others prefer proven breeders, thus eliminating the expense of feeding the animals until they mature.

We always preferred to buy young stock that was at

least three months of age, but not more than five months. In this manner, we were spared a good portion of the expense of feeding the rabbit until it was ready to produce, and it would have already shown whether or not it would develop into a well-proportioned animal for meat production. By buying rabbits in the three- to five-months' age range, we had ample time to become acquainted with the rabbit and its individual personality quirks, if any.

Proven stock is normally quite expensive, if the animal is a really good producer. Beware a "proven" doe which is offered at a "good price." It's all too often someone's way of getting at least part of his money back on a poor or marginal producer. For a really fine, proven production doe, expect to pay a pretty steep price.

While registration is not necessary in a meat production herd, all experienced breeders, large or small, advise the beginner to obtain a pedigree for each rabbit purchased. The pedigree does not mean that the rabbit is registered with the American Rabbit Breeders' Association, nor that the rabbit is even registerable. All a pedigree does is to give the family lineage, and that is all it is supposed to do.

It is a good idea to keep a family record on each and every rabbit bought or raised as a breeder. This enables the raiser to work out the best breeding program to ensure a good reliable meat source. Inbreeding and line breeding are often practiced in both commercial and show herds. Thus, if all stock is purchased from one rabbitry, the pedigree will show any close breeding, enabling the new breeder to plan his own schedule to prevent too-close inbreeding in future generations.

Chapter 3

Breeders

Handling Stock

Almost everyone has seen a picture, at one time or another, of a person holding a rabbit by the ears. This may make for good comics, but it makes for poor rabbits. It hurts the rabbit terribly and makes it afraid of the handler—it will be either excessively timid or overly aggressive toward him. So the first thing you must learn is how to handle rabbits.

The proper way to pick up a mature rabbit is to take up a handful of the loose hide and fur over the shoulders and neck of the rabbit and lift gently, sliding the other hand down under the rump to support the rabbit's weight when it is lifted. Good handlers never carry a doe or buck with its full weight hanging from the hide over the shoulders.

A method that we found useful in carrying a kicker is to catch up the handful of loose hide over the shoulders, then instead of putting the other hand under the rump, slide the free hand back under the belly to just in front of the back legs. The thumb and fingers press lightly against the front of the thighs, discouraging any wild kicking on the part of the rabbit. This method works especially well with does, which are usually heavier than the bucks.

Many people advise catching the ears of a fryer, along with the loose skin, when handling the younger rabbits. For a while I used this method, but discovered that the little rabbits could still scratch just as fiercely as if their ears were left free. (Holding the ears was supposed to somehow prevent clawing!) After accidentally breaking down the ears of a beautiful young doe when my grip slipped, leaving me holding only the ears, I changed my fryer-holding method to leave the ears loose. With the ears free, if my grip gave

A

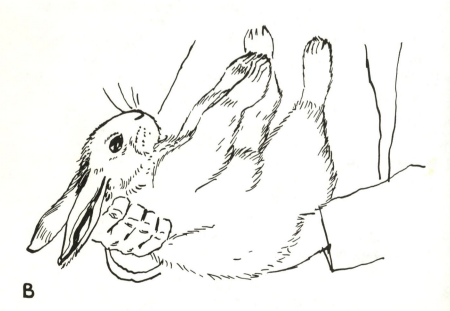

B

To inspect a rabbit, grasp the loose skin over the shoulders in the right hand and pick the rabbit up (A). With the left hand, cradle the rump and turn the rabbit feet up (B). Then, cradle the rump in the crook of the right arm, freeing the left hand to manipulate the rabbit during the examination.

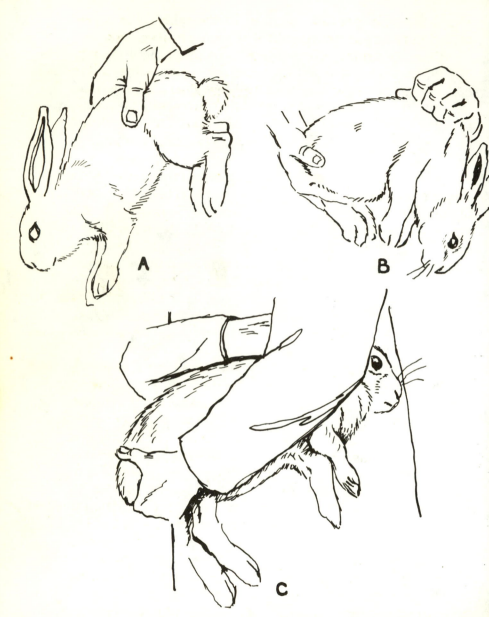

The proper way to lift and carry a rabbit weighing
less than five pounds is shown in A. If the rabbit is
heavier than five pounds, it should be lifted as in B
and carried as in C.

away, the only possible damage to the rabbit was a few pulled-out hairs as it slipped free. No more broken ears.

Does are usually the most easily handled of all the rabbits in the herd. Now and then, one will come across a doe that is high-strung and hard to handle, but as a rule they are fairly docile. Bucks are often great nuisances when it comes to having their share of everyday attention, but they resent being handled or carried, as when moving from one cage to another.

Never let a buck know that you are afraid of him. Like dogs, rabbits sense fear in a human and will take advantage of that fear. When a buck must be handled, do it quickly and aggressively. Reach into the cage and take the buck by the back of the neck and shoulders and quickly slide the hand down over the rump, so that the fingers are cupped around the hips and upper thighs. Hold on snugly but not tightly enough to make the buck uncomfortable. If he feels he is safe from being dropped, and doesn't feel pinched or cramped, he is likelier to ride along with you without causing trouble.

As with many other animals, talking to them seems to give them a sense of safety when being handled. It doesn't really matter what you say, but talk in a calm, quiet manner. This works not only with the mature breeding animals but with those young meat fryers, as well. Occasionally a person will be faced with a rabbit that has a wild streak. When this happens and the rabbit shows no signs of calming down or becoming accustomed to being handled, it shouldn't be saved as a breeder. The wild, nervous rabbit is very likely to become more so when the time comes for it to kindle and care for young. The easily upset doe, while she may never be aggressive enough to actually attack the raiser, may kill more than a few of her babies while jumping in and out of the nest to keep away from the handler.

I don't believe in making fetishes of production rabbits, becoming so attached to them that it is impossible to dispose of them when they can no longer function. But, I do think it is only humane to give a top producing animal the same care and consideration one would give a fine milking goat or cow or willing horse that does its full share of hard work. Good rabbits also work hard to produce meat for their owner and deserve the best care that owner is able to provide for them.

On the other hand, a poor producer is a drag on the fine animals in the herd. A doe or buck that simply cannot fulfill its designated role of a production unit causes the production of other animals in the herd to be more expensive. The doe raising only four or five fryers will eat just as much as the doe raising eight or nine. A buck that causes several does to miss at two or more periods during the year costs the owner just as much to feed as the buck that gets a nice litter with every service.

Another production cost to consider is the doe that is a poor milker. The saying that "the fryer is made in the nest box" is quite literally true. If a doe has a poor milk supply, then her young are going to take longer to reach slaughter weight than the litter of the doe having ample milk. The first three weeks are critical in the growth of young rabbits. And this is the period when they will be almost entirely dependent on the mother's milk.

Daily Care

Proper daily care of good breeding rabbits entails a little more than just dumping in food and providing water for them to drink. However, tending them on a daily basis is not a time-consuming matter. It can be done *during* feeding or watering. All it takes is a sharp eye to detect anything that appears unusual.

An unusual event in the rabbitry would be: a rabbit that hasn't eaten all of its food; one that appears to be listless, not perking up when the feed bucket comes near its cage; a bit of wetness around the mouth which may signify that the rabbit is too hot, has a bad tooth, or may be coming down with a cold; a rabbit that sits hunched up in an awkward-looking position that may mean it is getting a sore on its foot or the beginnings of some digestive problem; an animal that moves about with a limp, which may not be a serious matter but could signify an injured shoulder, hip, or foot.

When does must be bred, it is a good idea to check them for physical well-being. Make sure there is no wetness around the nose and mouth. Pick the doe up by the loose hide over the shoulders and neck as previously described. Rest the doe's back against your thighs, and with the other hand, press in on either side of the vent (external sex organ). If the color of the vent is pale pink or white, there is little use in taking the doe to a buck, as she is likely to refuse service. The vent should be reddish or a purple color for ease in breeding.

Check the ears for signs of ear canker (mites). Ear canker is not a disease in itself. It is caused by microscopic mites burrowing into the tender skin inside the ear. But, the mites can be transferred from one rabbit to another, either by a cage-to-cage migration or by physical contact while breeding.

Over the years, I discovered a method of determining the best of health in a breeding rabbit quite by accident. In handling and caring for the occasional sick rabbit, I found that almost without exception, the ill rabbit's ears felt cool or cold to the touch. Taking the discovery a step further, I started checking the ears of does due to be bred. If a doe's ears felt cold, she was allowed to wait another day or so before being taken to the buck's cage for service, even though she appeared in other ways to be perfectly healthy.

Medications, Their Use and Misuse

More than one would-be breeder has unwittingly ruined his chances of having a top-producing rabbit herd, simply by medicating them to death. Normally, rabbits rarely need any sort of medication other than a coccidiostat—a preventive medication used to control the coccus germ which is present in virtually every rabbit herd. Many medications available for use in rabbit herds today can do as much damage when improperly used as they can do good when they are truly needed.

Other raisers and I have known of people who went into their rabbitries each and every day with a hypodermic needle, and indiscriminately gave shots of antibiotics to any rabbit that wasn't fully alert—on its toes, so to speak. Soon the shot-giver would be "out of the rabbit business," never realizing that the antibiotics he'd given so freely were the basis of most of his problems with misconceptions, refusal to breed by does, sterility of bucks, and so forth. Invariably, these people blame bad stock for their failure.

A good motto for the beginner is one that is used by longtime breeders: "The best medication is the least medication." Learning to recognize the external signs of an impending illness in a rabbit is one of the most important lessons the new breeder can get. Most rabbit ailments are preceded by some sort of indication that all is not right. Only occasionally will some severe disease attack without prior warning.

Certain antibiotics, and not all of these are known as yet, can cause changes in the vaginal secretions of the doe, which in turn can result in a sterile condition. Some experienced breeders in recent years have reached the conclusion that some antibiotics, given too often or for too long a period, can cause blindness.

On occasion, of course, every raiser must use medications to stop an ailment that afflicts a fine breeding animal. If a heavy nursing doe, usually newly kindled, catches a minor cold, it is usually wise to give the doe an injection of penicillin. As a rule, if the raiser watches the herd closely for any symptoms, one shot is sufficient. If the injection is given early enough in the progress of the cold, the doe should recover by the following day. Giving a second injection the following day is safe. But the shots should not be continued for more than three consecutive days. The same thing holds true for a working buck that shows symptoms of having a cold.

Water-soluble medications for rabbits are among the most often abused. I have known so-called rabbit raisers who kept medications in their rabbits' drinking water on a continual basis, day in and day out.

Continual use of many medications causes not only upsets in the rabbits' systems; it can also create a situation where the rabbit becomes immune to the beneficial qualities of the treatment. Anything medicinal should be used *only* when the raiser has satisfied himself that the treatment is really necessary.

The one exception to the above would be the use of a good coccidiostat. This medication prevents a multiplication of the coccus germ which is apparently present in all rabbits. It is given on a regularly spaced program for the best results; when dosage is stopped for any length of time, young rabbits become infested by the oocytes (eggs) of these internal parasites, causing problems with diarrhea, poor growth rates, and even death in severe cases.

A coccidiosis medication, like all others, should not be used on a daily schedule, however. It is used most effectively when added to drinking water for three consecutive days out of each thirty-day period. On rare occasions, in a herd whose health has been ignored for some time, it may become necessary to administer a heavier dosage than that suggested on the package label, and over a longer period than the normal three days per month. If such should be the case, increased dosage and time should be discussed with a veterinarian or at least the field representative of a good feed company.

Sexing and Breeding Procedures

Is it a female rabbit or is it a male rabbit? How do you tell? Once you have worked with rabbits for a while, you'll find it usually very easy to determine. I say easy advisedly; I know people who have raised rabbits for twenty or thirty

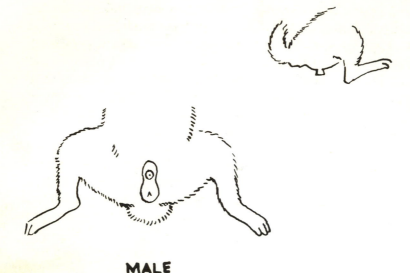

MALE

To sex a small rabbit, lay the rabbit on its back with its head toward you. Grasp a hind leg between each thumb and index finger and use the thumbs to manipulate the external sex organs. As can be seen, the difference between the male and the female is subtle.

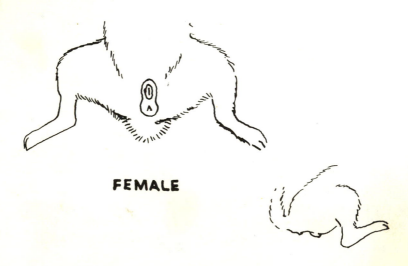

FEMALE

years and who still make an occasional mistake in sexing young rabbits. I've made a few of those mistakes myself. In fact, one of the finest herd bucks we ever raised was originally put into a young doe cage. I was so sure the rabbit was a doe that it wasn't until he started spraying urine all over adjoining doe cages (and me!) that I finally discovered he was a sure-enough male rabbit. The buck was a good three months old when I discovered his true sex.

It is rarely necessary to determine the sex of a rabbit less than six or eight weeks of age. Show breeders sometimes want to know the sex of an individual litter, so that the better babies can be given special care. In this manner, the rabbit is assured the best possible chance of growing out to become a winner on the show table.

Should you desire to know the sex of a special, very young bunny, find out by holding the little rabbit in your left hand (lefties, reverse procedure), *upside down*, with the fingers holding the legs to prevent squirming and kicking. With the index and middle finger of the other hand, press in gently on either side of the vent (the sex organs) area.

If the little rabbit is a buck, a minute tubular protrusion will be noticed. If the bunny is a doe, there will be an almost equally minute slit, instead of the protrusion noted on the male. Once the rabbit is around six to eight weeks old, the tiny protrusion becomes recognizable as the sheath or foreskin; the penis will by this age push out through the little protrusion when pressure is applied on either side. As a rule, the testicles of a male rabbit don't become really visible until he reaches the six- to eight-weeks' age range. Some males don't have noticeable testicles until they are even older.

The medium-sized breeds of rabbits are not fully mature until they reach 5½ to 6 months of age, as a rule. Of course, there are exceptions to this. You may now and then find a rabbit that is mature at less than 5 months of age, but breeding that young is not advised.

Medium-sized does are usually bred for the first time at about 5½ months, which results in the first litter born when the doe is 6½ months. Bucks are ordinarily started to work at about 6 months of age. Young does are usually first worked with an older, experienced buck, but this is not really necessary if the raiser has only a young buck. The young, inexperienced buck is usually first worked with an older doe, too. But again, this is not absolutely necessary. Rabbits, young or old, will sooner or later do what they're expected to do. Working with young, inexperienced stock, you may just have to wait longer for the breeding to come about than if one of the participants is an experienced breeding rabbit.

Mating: For the best results on breeding day, always take the doe to the buck's cage. Does resent intruders into their cages and may fight a buck on their own territory. Bucks are extremely snoopy, and when set in a doe's cage, may waste considerable time just poking around the cage and showing little interest in the doe.

Before taking a doe to a buck's cage, check to see if she is ready to breed. Rabbit does are not supposed to have a regular estrus (heat cycle), but if a doe is taken to a buck when she is not in the mood for breeding, the raiser is wasting his time. When the doe is ready for mating, the vent will have a slightly swollen appearance and look a little red or purplish in color. It is ordinarily useless to take a doe to the buck when her vent is a very pale pink or whitish color. In some cases the doe will fight the buck, possibly causing harm. In most instances, the doe will huddle back into a corner and squeak or whine, and no buck can possibly dig her out of her chosen hiding place. If the buck becomes too insistent and aggressive, the unwilling doe will try to run away from him, or may even resort to fighting to prevent him from mounting.

When the doe accepts service by the buck, the buck usually falls over to one side, or sometimes backwards. Nothing to worry about. Within a minute or so, he will be up and ready to repeat the mating if the doe is willing.

Occasionally, one may come across a screamer. That is, a few bucks don't fall over when they ejaculate; instead they stand up straight on their back feet and let out a piercing scream. No one knows exactly why a buck acts in this manner. Some breeders figure they experience pain with ejaculation. Others believe these bucks simply enjoy their work. I suspect the latter is true, since the screamer is always ready to repeat the action within a minute or so. And usually screams again!

The doe should not be left unattended in the buck's cage, even though she may be willing and eager for mating. Once the doe has been bred, she sometimes grows impatient with the buck's continued advances. An angry doe can severely injure a buck, in some cases even partially castrating him. Any injury, severe or not, may cause the buck to be a little afraid of an aggressive doe in the future. A buck that has been injured by a doe can become so timid with future does that he won't even try to mate.

All stock saved for future breeding should be separated from the opposite sex when the rabbits are no older than 2 months. Some raisers have found, to their dismay, that there are extremely precocious young rabbits able at the age of 2½ months to become pregnant or to fertilize the ova of other young rabbits. More than one little 3½-month-old doe has been lost because she was left in the same cage with her littermate brothers too long! Usually not only the little doe, but her too-early litter are all lost. If the doe survives, her litter is ordinarily lost, since the doe is confused and doesn't know yet how to take care of the little ones. In some instances of too-early pregnancies, the doe may survive the kindling, then die two or three days later from some invisible internal damage.

Pregnancy: Young does may be kept in a community cage until they are about four months of age. After this, they should each be housed in a single cage. Young does will often ride each other, with an aggressor doe acting the part of a buck in mock matings. The doe that is ridden may ovulate and experience what is called a false pregnancy. She goes through all the motions of being pregnant and refuses to breed with a buck; she may even gain some weight. The false pregnancy normally lasts only about seventeen days, at the end of which the young doe usually goes through the motions of making a nest for young, even pulling hair to bed them.

Shortly after the false pregnancy ends, the doe is likely willing to breed. If the doe has been bred, she must be moved to another cage, preferably before she is fourteen days pregnant. Later moving may make the doe upset, perhaps even causing her to reject her babies when they are born.

Baby rabbits are born about thirty-one days after the doe is bred, so on the twenty-eighth day, she should be given a nest box. Make certain the box is clean. Use of a disinfectant in the box helps to insure health in the tiny rabbits.

For nesting material use oat straw; shredded cane stalks (bagasse); clean, dry wood shavings; clean, unsprayed grass hays; and even excelsior. If the shredded cane stalks are used, they should be placed in the bottom of the box with a thick layer of straw or hay over them. Bagasse tends to pack into the lower part of the box, which makes an excellent insulation for winter nests. The straw or hay on top of the cane stalks acts as the actual bedding—sort of like a solid mattress with a feather comforter on top!

The expectant doe should be given some extra hay or straw to satisfy her packing instinct. Most does are very industrious in building nests and apparently love to carry nesting material in and out of the box in preparation for their litters. Make a daily check (more frequent checks are often advisable) on the amount of straw the doe still has in the box. Sometimes quite a lot of straw is lost through the cage floor during nest building.

For the summer nest, straw or clean wood shavings are usually sufficient. Many good does start pulling some of their hair a day or so before their litters are due. By the time the litter is born, the doe's belly and sides may appear quite bereft of hair. No cause for worry, since the hair pulling doesn't hurt the doe, and it serves two purposes: bedding to

A newborn litter in its nest is protected by an ample supply of fur the mother has pulled from her own body.

keep the naked newborns warm, and baring the nipples so that the new babies can nurse easily. Quite often, just before a doe is due to kindle, she may give herself a very thorough bath, paying special attention to her belly and feet.

Kindling: The doe should be left alone when her kindling time nears. Does, even those old experienced ones, may become very nervous during kindling if someone is fooling around the cage. Just wait and be patient; in time, nature will take care of the situation and you can count those little rabbits!

After the doe has finished delivering her young, look into the nest briefly to see if she has covered them well with nesting material and fur. If so, leave them alone until the next day. If the babies are not covered well, it is possible to hold the doe by the loose hide over the neck and shoulders and pull some fur from around the nipples to bed the babies. Often, this manual pulling of fur will remind the doe that she still has work to do, and she will proceed to pull more fur after you've stopped. The hair on the belly pulls very easily just after the doe kindles, so you're really not hurting the doe.

Now, give the doe at least several hours, preferably overnight, to quiet down after delivering young. Next day, go into the rabbitry and give the newly kindled doe a special treat such as a large comfrey leaf, some oat hay, or a carrot. Usually this temporarily takes the doe's mind off her nest while you look over the babies and count them. If the doe seems nervous about you looking over her young, sometimes a gentle handrub back over the ears calms her fears. Never allow yourself to become cross with a doe that seems nervous and fusses at you for looking in on her litter; she is only doing what nature dictates—protecting her young. Work with her calmly, showing that you don't mean her or her litter any harm, and she will usually be more amenable the next time you check her litter.

Post-Natal Care: New babies should be checked just about every day until they start coming out of the nest. Sometimes, a new little rabbit will succumb to some mysterious ailment or injury overnight. Left in the nest by a careless raiser, it deteriorates rapidly and dies, especially in warm weather. Since the other babies can't leave the nest, and the doe rarely if ever removes a dead baby from the nest, this exposes the remaining littermates to disease germs, flies and their contamination, and even to possible injury from maggots.

Small rabbits usually won't leave the nest until they are about three weeks of age, unless the doe is short of milk. The occasional bunny that may be found sitting outside the nest with its mother should be returned to the nest. It may have come out on its own, but it may have been pulled out of the nest by the doe. A little rabbit will sometimes hang on to a nipple and be pulled from the nest when the doe is ready to stop nursing.

Baby rabbits' eyes should be open by the time they are twelve to fourteen days old. If a bunny still has closed eyes after this, there may be infection in the eye. This can come about from minute particles of dust in the nesting material which get into the bunny's eye, or it may happen when the doe is not as clean as she should be with her nest and wets in it.

To open the eye that is matted shut with infection, hold the little rabbit securely with one hand, then with the thumb and forefinger press outward (very, very gently!) on either side of the eye slit. When the slit is partially open, wipe the eye clean with a piece of absorbent cotton. Then apply a small amount of an antibiotic eye ointment, or drop just one drop of ordinary eyewash into the eye. Plain old Murine® will do, or any other gentle eyewash made for human use. The infected eye may need to be reopened and cleaned several days in a row. If the eye is opened before infection becomes drastic, though, the eye will be normal, with no impairment of sight.

When the babies start coming out of the nest and sampling the foods provided for the doe, very gradually add more feed to the doe's ration. You may also want to provide some oatmeal or oatmeal groats for the special benefit of the youngsters. It doesn't harm the doe if she eats a little of the oatmeal or groats along with the babies. By the time the rabbits are about a month old, they often are seen at the doe's feeder eating bits of her ration. They don't eat very much at a time, but nibble frequently. The doe still furnishes them with a considerable amount of milk, too.

During the two weeks between one month and six weeks of age, the young rabbits increase their intake of solid food substantially. Increase the feed amounts to keep up with the extra demand, but never in such quantities that there is much feed left over. The animals should receive fresh feed each day, which encourages hearty eating and good growth rates. During this time, water is of utmost importance. Growing fryers and their nursing mother should never be allowed to run out of water.

When weather is very hot, the nest box may be removed when the little rabbits are from five to six weeks old. Some

The cute little bunnies remain with the mother until they are fully weaned.

raisers in areas where high temperatures are quite stable in the summer remove the box even at one month. During winter months, keep the nest box in the cage until the bunnies are about six weeks old. The little rabbits will retire to the box when temperatures are very cold; it provides them with extra warmth. You may have to provide a clean nest box with a small amount of clean bedding when the box is left in the cage for those six weeks.

When the young fryers are about six weeks old, the doe may be rebred. First, check the doe for physical condition. She should have a flabby belly after nursing a litter for six weeks, but overall, she should not appear miserably thin and underfed. If the doe is in good condition, she may be

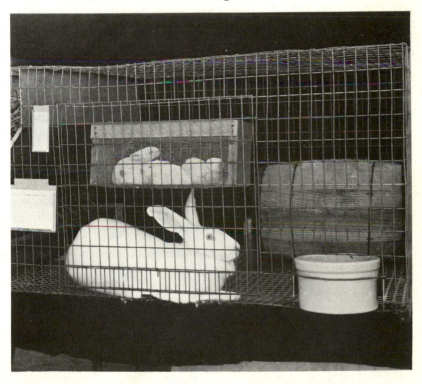

During extremely hot weather, some breeders place the bunnies in a homemade "cooling basket," hung in the hutch in view of the mother, but away from her body's heat.

taken to the buck's cage for rebreeding. After mating, she is returned to her own cage and litter. Feed consumption by the litter increases rapidly after the doe is rebred, since her milk supply rapidly dwindles after conception.

If, however, the doe is terribly thin and "nursed down," wean all but one or two of the litter before she is rebred, and keep her on a full ration of feed for a few days. The one or two bunnies left with her prevent milk from building up in the breasts and caking.

With this type of care, the doe that has nursed down should start gaining a little weight within a week. Just a bit of livestock calcium dissolved in her drinking water may replace the depleted calcium supplies in her system. Extra vitamins are also useful in building a nursed-down doe back to her usual fitness. (Adequate vitamins in routine feeding of a nursing doe can prevent much of the nursing down that comes from feeding a large hungry litter.)

If you left one or two fryers with the doe while she regained her good condition for breeding, wean these fryers within two weeks after the doe is rebred. It is usually possible to rebreed a doe within a week or ten days after the major part of the litter is weaned, since she will have probably gained enough weight by this time to ensure her staying in good health.

Sometimes in weaning fryers from the doe, you may notice a bit of weight loss in the fryers those first few days in a strange cage. This weight loss can be minimized in either of two ways: first, move the doe to a different, clean cage, leaving the little rabbits in their own home cage. This reduces whatever trauma they may suffer at the loss of the mother. Second, instead of moving the young rabbits into a strange cage with strange new fixtures as well, transfer the doe's feeder, water bowl (if one is used), and also the gnawing block (if one is used to keep teeth ground down properly).

Having the mother's scent along with them in their new home minimizes the shock of losing the mother. When the fryers are first moved into a new cage, giving them a large handful of clean straw will also take their minds off being moved. They'll be more interested in eating than in noticing their strange quarters!

Special Situations: During late pregnancy in hot weather, take care to prevent does from becoming overheated. If the weather is very hot and temperatures hold above 80 degrees Fahrenheit, the doe that is near kindling can be placed on the ground, preferably inside a building or enclosure where other animals cannot disturb her. Just set the doe, along with her prepared nest box, down on the cool ground inside of a floorless wire "cage."

The doe's resting place should be back out of the way, so that she won't be unduly disturbed during labor and kindling. When she feels the cool soil beneath her, she usually lies down on her belly until the young start arriving. After delivering and feeding her babies in the nest box, she normally goes back to the cool ground and lies there resting for some time. If the doe is highly nervous and upset, she can be left on the ground for a few hours, until nightfall when the temperature usually lowers.

When returning the doe to a regular cage, along with her nest box and young, always return her to the cage she has lived in previously. Changing her to a strange cage at this time could cause her to disown the young, or to eat them, which is called cannibalism. Don't clean or change the scent in her cage in any way while she is on the ground. Be sure to provide her with a small bowl of fresh cool water. Some does won't drink while in labor; others appear to be quite thirsty.

Immediately after the doe kindles, give her a little feed; you may also give a small amount (several stalks) of good, clean oat hay if it's available, or a large comfrey leaf. Or try instead some freshly pulled clover, clover hay, or a small handful of alfalfa hay. Plantain, about four or five leaves, seems to be thoroughly enjoyed by newly delivered does and has a beneficial effect.

Some people advocate actually *cutting back* on a freshly kindled doe's food for two or three days. I could never see the logic behind this. The doe's system undergoes a strain for a few critical days in delivering young and immediately making milk to feed them. If the feed is cut back, what does the doe's system manufacture milk from—fresh water and air?

We always had excellent results with gradually *increasing* the mother's ration immediately after she kindled. If she normally received six ounces of feed, plus a little supplemental feeding, her ration immediately after delivering young was increased by three or four ounces. Over the next three or four days, her ration was increased to *whatever amount she would clean up in a twenty-four-hour period*.

First-litter does were always fed very carefully to find out their digestive reaction to the extra feed. A few does will have soft droppings if the feed is increased too fast. Most does, however, react favorably to having a comfortably full belly when their bodies are converting most of their feed intake into milk. Once we were thoroughly acquainted with a doe's reaction to the additional feed, we sometimes gave a full ration as soon as we discovered that she had newborns to feed.

Getting Does Bred

Normally, there is little problem in getting rabbits to breed, if you are sure to check the does for readiness. When a doe has large fryers with her and is unwilling to breed, usually there is a good reason for her reluctance. She may be out of condition—nursed down a bit more than is good for her. She may be too fat. If so, cut her ration of feed down for several days until she starts showing a slight weight loss. Should the doe be too thin, provide a small extra ration of feed for a few days. It isn't usually necessary to give the doe extra feed for more than one week. (The nursing doe is normally the only one that needs extra feeding to regain lost weight.) If a doe is unwilling to breed, and is not too fat, nor sick, it often helps to house her right next door to a buck. In some cases, a bit of extra protein increases the doe's interest in breeding. More often, an ample supply of vitamin E enhances her willingness to accept the buck.

The young doe ready for her first mating may be a little frightened of the buck on her first visit to his cage. If so, she should not be forced to accept him. Return her to her own cage and give her a bit of attention, talking quietly to her. Chances are that the next day the young doe will be willing to accept the buck, conceiving a nice first litter.

On the other hand, there exists the aggressive doe, usually one that has mated successfully on one or more occasions. Some does try to ride the buck, and a few become even very aggressive, biting around the buck's ears and loin. As a rule, bucks won't tolerate this sort of action for very long, though some seem to enjoy the aggressive doe's advances for a while before asserting themselves.

At times it is the buck that is slow or even completely uninterested in breeding. If he is healthy, the raiser can sometimes increase the buck's desire by little subterfuges. A good massage along the back has perked up more than one buck's interest in breeding. One breeder I know suggested that I *tickle* a buck that shows little interest in the doe! It actually works in many cases, too. Just rub the buck's back, then work the fingertips down to the sides just in front of the back legs, and tickle his sides and thighs. The buck will likely start flinching and making little skipping motions with his back feet. Then, just set the doe into the buck's cage, and quite often he proceeds with the mating on his own. Tickling and rubbing under the chin sometimes livens up a sluggish buck. If the buck is entirely too fat for successful breeding, it is necessary to put him on a strictly rationed diet for several days before trying to mate him with does.

Very few people practice what is called forced breeding in their rabbit herds. This is simply holding the doe in the proper position for mating so that the buck can mount and breed her. The reason so few successful rabbit raisers use this forced breeding is its low rate of conception. In a herd of good healthy animals, the natural rate of conception is normally at least 80 percent. With the use of forced matings, the conception rate is rarely higher than 20 percent, which is a waste of the breeder's time and effort. It is also a waste of herd time, since pregnancy can't be determined until approximately two weeks after the forced breeding.

Determining Pregnancy

Most commercial rabbit breeders use a method called palpating to determine whether or not a doe is pregnant. This is done by holding the doe by the loose hide over the neck and shoulders so that she is lying flat on her belly. Then, with the other hand, feel the belly for small marblelike lumps which are the developing embryos.

To me, palpating is rather a difficult (and uncertain) method of finding out if the doe is pregnant. Not only have many so-called experts at palpating misjudged pregnancy or nonpregnancy, but I have always wondered if some of the birth defects in rabbits couldn't have been traced directly back to inexperienced people exerting too much pressure on the doe's belly. All that gouging around in the doe's belly couldn't possibly benefit the young in any way, in my opinion.

The method we always used, after I discovered how hard it was to determine the presence of young in the uterus by palpating, was simply to return the doe to the buck's cage fourteen days after the original mating. This method is called unreliable by some raisers. But in our years of raising rabbits, working with several hundred does and numerous bucks, we had exactly *one* doe accept service on this second trip to the buck's cage and then kindle a litter just fourteen

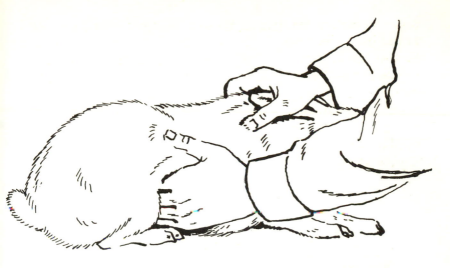

Palpation is a tricky procedure. Using the right hand, restrain the doe by holding the loose skin over the shoulders and the ears (to restrain the head). Thrust the left hand under the rabbit, between the hind legs slightly in front of the pelvis. The thumb should be on the right side, the fingers on the left. The idea is to feel for the tiny developing babies, but they are often difficult to distinguish from the rabbit's organs. And if the doe is at all tense, the procedure is almost impossible and injury may result.

days after the second service. Almost without exception, the already pregnant doe tries to run away from the buck or hide in a cage corner. Frequently the pregnant doe makes a grunting or growling sound when returned to the buck for this test mating, and many does fight at the buck. If your doe fights, remove her from the buck's cage quickly, or she may hurt him in some way.

Even if the doe does accept a second service by the buck fourteen days after the first mating, it is still wise to furnish her with a nest box and nesting material fourteen days after the second mating (twenty-eight days after the first service). This is just in case she is one of those rare does that accepts the buck even though she is already pregnant from the first mating.

Multiple Matings

Experienced raisers advise the beginner to never leave the doe in the buck's cage for more than a few minutes. The reason most often mentioned is that the angry doe can do real physical harm to the buck.

Another, less publicized reason for not leaving the doe with the buck for more than a few minutes is multiple litters. A rabbit doe can conceive from just one of her two ovaries when she is bred. This one ovary may produce a small litter of only three or four, but it may produce a good litter of seven or eight! If the doe is left with the buck for a long time, she may accept one service during the first few minutes and conceive from that mating, but from only one ovary. Left with the buck, she may accept a second service as much as two or three days later, conceiving a second litter from the other ovary. In which case, the first litter kindled does not receive a full supply of milk, since the doe's system is diverting a portion of its food intake to the developing embryos of the second litter, yet to be kindled.

Then, when the second litter conceived is finally kindled, they are usually lost, being smaller and weaker than the first litter and unable to compete with them for milk. Also, the doe may try to build a second nest, usually outside of the nest box, which means the second litter delivered will be lost due to exposure.

To prevent damage to the buck, and to prevent multiple litters being born several days apart, most experienced raisers take the doe (especially the young and inexperienced one) to the buck early in the morning for a first mating. From six to eight hours later, the doe is returned to the buck's cage for a second mating. In this way, if the doe conceives separately from the two breedings, both litters will be kindled fairly close to the same time, usually within half an hour of each other.

Fostering of Young

Many does have more than a normal litter of eight. When this happens, often a part of the litter is lost, since the average doe has eight nipples. Or if all of the babies are strong and aggressive, pushing for their share of milk, some of the litter members will likely be stunted, because the weaker ones will get only what milk is left after the stronger bunnies have had their fill. The runts are more susceptible to infection than those that are strong enough to get their fill of good milk.

Therefore, it is wise to breed two or more does on the same day so that they kindle at the same time or within a few hours of each other. If one of the does has an exceptionally large litter (more than nine or ten), a part of it can be farmed out to a doe with a smaller litter.

Farming out means getting one doe to act as a foster mother to part of another doe's young. The majority of good does will adopt babies if their transfer is handled properly. Some people advocate the use of some sort of oil or powder on the nose of the adopting doe, so that she doesn't smell the strange babies before they have acquired the scent from her own nest and litter.

A scented oil works, but the use of talcum on the doe's nose is questionable.

Most does (if they are amenable to adopting extra babies at all) can be given spare babies in a much simpler fashion. Take any extra babies from the nest of a too-large litter. Remove all of the hair and nesting material from them. Take the adopting doe a special treat like clover, carrot, comfrey leaf, or a handful of hay or straw that is different from what she is accustomed to eating every day. Put the special treat into the back of her cage, away from the door so that she is out of the way of the nest box when you pull it forward. Put the strange babies under the doe's natural litter, then replace the fur over them. Give the adopting doe a little rub over her ears and back, which gives her some assurance that you mean no harm. By the time she has finished her treat, the extra babies now in her nest have acquired her scent from her own babies and the nest. It is a rare doe that won't adopt outside bunnies in this manner, though the possibility exists that a doe won't adopt under any circumstances.

Breeding for Stamina

In virtually every rabbitry, there are one or more rabbits that just never seem to become ill, no matter what. A certain doe never has any problems in kindling large litters; she apparently manages beastly weather without undue stress; and her daughters seem able to carry on in their dam's footsteps. This is called stamina in the brood stock.

These hardy animals are the ones from which new young stock should be saved. The buck, too, passes stamina or a lack of hardiness on to his offspring. Sometimes, a rabbit with all these good traits does not pass stamina along to its young.

To set stamina in a herd, most breeders use what is called progeny testing or projection testing. This is saving stock from only those rabbits that have no problems under normal circumstances. The young stock is then worked for at least three litters. (Some commercial breeders prefer using a five-litter trial period.) Records are kept on the younger stock to see if they are able to produce similar to or better than their parents.

One of the best ways to set stamina in a herd is by line breeding. This method breeds uncles to nieces, aunts to nephews, cousins (not of the first-cousin generation), grandsire or granddam to granddaughter or grandson. Use only the very best two or three young does and bucks in a litter as future breeders. If one of them doesn't pass on hardiness to its offspring, this rabbit is culled from the herd. (Yes, adult rabbits are excellent for cooking.)

Breeding dam to son, or sire to daughter, can produce some very good rabbits. On the other hand, inbreeding isn't usually advised for the novice breeder, since the beginner has not had time to learn the genetic traits of his basic stock. Breeding sister to brother is literally taboo, except for testing to find out if there are bad genes being carried in a family line. No stock should ever be saved from a brother-sister mating.

If you start out with only two rabbits, a buck and a doe, you can find out quickly just which rabbit throws the greater amount of stamina to offspring. When the pair's first litter reaches weaning age, save the very finest-looking and healthiest-appearing two little does and one buck. When the young rabbits reach maturity (about 5½ months for the medium-sized breeds, four months for the smaller breeds, and twelve to fourteen months for the Giants), the two young does are bred back to their own sire, and the young buck is bred to his dam.

Watch the resulting litters from these inbreedings very carefully, observing general health during the fast-growth period up to two months. Note also any physical faults such as weak, crooked legs, droopy ears, or pinched-in hips.

As a rule, one of the young does does a better job of raising out her litter than her sister. But if both of the does do an equally good job of milking and finishing out their young, it is safe to say that the original buck (sire of both the young does and their litters) is throwing a considerable amount of strength and hardiness to his offspring. If he were passing on any sort of weakness, it would become more apparent in the litter out of his own daughter.

The same is true of the litter resulting from the original doe being bred to her own son. If all three inbred litters grow out equally well, future breeding stock may be saved from all. Chances are, however, that one of the litters will stay healthier and grow out faster than the other two. This is the litter from which to save stock, but save only the finest two or three rabbits.

Once you become firmly acquainted with the traits your stock possesses, you can carry the inbreeding even closer, such as half-brother and half-sister breedings, to set some especially desirable trait into the family line. Offspring of such a mating should not be bred back to their own dam or sire. Instead, the resulting stock should be bred to a more distant relative, or even to an unrelated rabbit of an equally good family line. Inbreeding, no matter how good the family lines, when carried too far and too close over too long a period, will weaken those finer traits the breeder is trying to set in the herd.

Chapter 4
Nutrition and Feeding

Nutritional deficiencies can yield some bizarre results. The crooked and splayed front legs resulted from a magnesium deficiency in this rabbit's diet.

The biggest continuing expense in rabbit raising is feed. Most commercial rabbitries purchase pelletized feed for their animals. The homesteader will probably want to avoid buying feed as much as possible, and, advice to the contrary notwithstanding, it is entirely possible, practical, and conceivably advisable to feed homegrown foodstuffs to the homestead rabbit herd.

There's a seductive ease to the use of pelletized feeds, purchased in sacks from the local feed mill. Feeding the rabbits this way requires no thought. It is a simple once-a-day chore.

For some rabbit fanciers—those with no time or space to grow feed themselves—commercial feeds are excellent. For the homesteader trying to be self-sufficient, it can be a bit contradictory to buy feed for the rabbits you are raising so that you don't have to buy meat.

In considering the production of rabbits and the use of grown and natural feeds, one must concern himself with the genetic structure of modern-day domestic stock. For decades, both commercial and show breeders have selected for future breeding only those rabbits which prospered best on concentrated pelletized feeds. Prior to the mass production of pelleted feeds, the rabbit breeder was forced to choose those rabbits that could produce and prosper on rations of natural grains, often grown by the breeder himself. After genetic selection of the stock that responded best to concentrated feeds, however, most modern rabbit stock did not do well on altogether unprocessed feeds until several generations had been selected for their abilities to produce well on natural feeds.

To obtain the best results in switching a rabbit herd from commercial to natural feeds, practice the reverse in genetic selection of stock. Instead of choosing those rabbits that respond well to only pelleted feeds, choose only those rabbits which have shown the best growth rates on diets of partly pelleted feeds, part natural grains, and other foods. Then gradually decrease the amount of pelleted feeds over several generations.

In time, a rabbit raiser could virtually discontinue the use of commercial feeds. Rabbits have been raised, and quite successfully, long before pelleted feeds were invented. But, just as we have lost some of the nutritional value of certain foods in the endless quest for bigger and better, the genetic traits which allowed rabbits to prosper on natural feeds have gradually been downgraded until today's domestic rabbit is genetically geared to respond only to the highly concentrated feedstuffs produced commercially. These original genetic traits must be reintroduced into the homestead rabbit family line if natural feeding is to be completely successful.

Since nutritional content of pelletized feeds may be lower than required by heavily producing rabbits, supplementary feeding is usually required. Supplements are used in virtually all successful rabbitries, though it may not be called supplemental feeding by those who use it most. For almost all commercial rabbitries, supplements are given not in the form of feeds, but in water-soluble vitamins and minerals through the drinking water. This is true in those rabbitries that go the complete pelletized route of feeding, as well as those that use pelletized feeds but also provide small treats of natural feeds such as garden produce. The concentrated pelletized feeds simply do not provide sufficient nutrients for best results in the rabbit herd.

Water

While water is not ordinarily considered a nutrient in itself, water consumption is an absolutely essential part of the rabbit's diet. Without water, conversion of feeds into health-giving nutrients is impaired. And insufficient water to drink decreases the rabbit's appetite, thus depleting the necessary amount of nutrients.

Consumption of large amounts of water enables the nursing doe to convert foodstuffs into milk for her young. A too-small water supply causes her milking abilities to deteriorate rapidly. And growing young rabbits must have ample water to convert foods into bone and muscle tissue.

A doe that is nursing an average litter often drinks as much as ½ gallon of water in a day. The doe and her litter that are eating some solid foods and still nursing frequently consume as much as a gallon of water in twenty-four hours.

Breeding bucks and dry does (unbred or pregnant but not lactating) don't drink so much, but they should be provided with at least a quart of fresh water per day. A one-pint bowl may be used for the single rabbit and filled twice daily, so that the rabbit doesn't wind up during the night hours without any water at all. Rabbits drink more at night than during the daytime.

The raiser should never add fresh water to a bowl or waterer partially filled with stale water. Each cage should be provided with completely fresh, clean water every day.

Protein

A shortage of protein in the diet can cause numerous problems in the rabbit herd. One of the first symptoms of protein deficiency is a lack of adequate milk production from does. Tiny babies in nests show the inadequacy of milk supply with lanky, empty little bellies and poor growth during the first days after birth. A well-fed newborn rabbit should look almost as if it has swallowed a miniature ball, its belly will be so full looking; and sometimes, the whitish color of milk can be seen through the thin wall of the tiny belly.

A second symptom of protein hunger is a tendency among does to eat a part or all of their litters during or immediately after kindling. Rabbits are not by nature carnivorous, but they will sometimes turn to cannibalism to try to get their required amounts of protein.

Another symptom of protein hunger can be noted in fur eating. In groups of young rabbits caged together, this may first be noticed in ragged-looking fur around their heads and faces. As a rule, the rabbit that is guilty of the fur chewing won't show chewed fur itself if it is caged in a family cage or a finishing pen. It is normally only the rabbit caged alone without access to other animals' hair that resorts to eating its own fur, leaving ragged, chewed-off places along its sides and hips.

Does with young in the nest may eat the fur which they pulled to cover their babies. Sometimes, the protein hunger doesn't become apparent until the doe's litter is older, already furred out. Then, all of the litter may show up with ragged, tacky looking fur, but the doe will probably not show signs of chewed fur herself unless her young also are starved for protein. Increasing the protein content of her feed ration will often stop the doe's fur eating.

Protein Sources: Most commonly used as a protein supplement is the pelleted Calf Manna®. Other sources may be more desirable in the homestead rabbitry, however. For example, comfrey contains almost all of the amino acids that are essential for high-quality protein. Legume hays are excellent sources of protein, but you may like to use comfrey leaves in addition to hay for milking does and small rabbits.

If one of the highly concentrated, pelleted high-protein supplements is used, small amounts of the legume hays may be used too, but with caution. Too much of a good thing (a too-high ration of protein to other ingredients) can cause digestive upsets in rabbits, particularly in the very young and the newly kindled doe. A too-high protein concentration can also cause a repressed growth rate in the young animals, rather than the efficient growth expected when the ration is kept balanced with other nutrients in the feed.

Overall, a ration that contains between 15 and 17 percent protein is considered best for the herd. Nursing does and very young rabbits (between three and eight weeks

of age) can make efficient use of greater percentage of protein, up to about 22 percent, with an average of 18 percent
preferred by most successful raisers. Working rations for
breeding bucks and dry does—also for growing stock past
two months of age—should have about 16 percent protein
content.

Vitamin Requirements

Domestic rabbits require vitamins E, K, A, D, and the Bs.
Vitamin C has never been proven necessary to a rabbit's
diet, though it does them no harm that has ever been noted.
Therefore, you could give rabbits pieces of citrus fruits or
peels and other foods containing vitamin C.

Vitamin E

Along with protein, vitamin E is one of the most important
nutrients required by rabbits. It affects the reproduction
abilities of every rabbit. A deficiency of vitamin E can cause
the following problems in the rabbitry: lack of interest in
breeding in both bucks and does; poor fertility among both
bucks and does; abortions and miscarriages by does.
Natural vitamin E has been proven by breeders to be more
reliable in its benefits to the herd than a synthetic form of
the same vitamin.

A vitamin E shortage in the herd's ration may be slow
in manifesting itself in these results. The trend toward
lower production may be so gradual that at first it is barely
discernible. But once the deficiency becomes entrenched in
the breeding stock, rebuilding the previous productivity is a
slow and costly process because of lost litters from conception failures, abortions, or miscarriages.

Vitamin E Sources: Whole cereal grains provide good quantities of vitamin E. Whole wheat is probably the most ap-

proved source of this vitamin. The wheat may be fed un-processed to mature rabbits. Husks on the grains may cause intestinal damage to the very young rabbit; therefore, steamed and rolled wheat is generally preferred for whole-herd feeding. Too, the rolled grain is used more efficiently by rabbits, since the animals do have a habit of wasting some food if husks are left on grains.

If steamed and rolled wheat is either unavailable or entirely too expensive, you may use wheat germ along with the other grains making up the rabbit's ration. A teaspoon or so of the wheat germ placed on top of the rabbit's daily food, takes the place of the whole wheat. Or, five or six drops of wheat germ oil (the kind available in veterinary supply de-partments of drugstores, usually in a bottle with a dropper) on top of the feed each day also supplies the breeding rabbit with sufficient vitamin E.

Vitamin K

This is one of the most difficult vitamins for which to de-termine a need in a herd. For instance, vitamin K helps to reduce the possibility of respiratory ailments among rab-bits, but there are so many other equally important, if not more important causes, that the novice breeder is hard put to decide if the problem stems from a lack of vitamin K or some other factor instead.

Colds (or *apparent colds*) can result from living in dirty, drafty quarters, or from ammonia, a by-product of ac-cumulated filth in the rabbitry. However, respiratory ail-ments and weaknesses are so often hereditary that you'd be often wiser to dispose of the rabbit that has colds frequently than to try to hang onto it and treat it, thereby risking infec-tion in some or all of the stronger animals. Though the rab-bit may react favorably to any one of several methods of pre-vention or cure, when the weakness is inherited the rabbit will likely catch a cold whenever it is subjected to any sort of stressful condition (as when a doe catches a cold each and every time she kindles a new litter). In such a case, the weaker animal is likely to pose a greater threat to the health of stronger animals than keeping it would be worth.

Vitamin K is obtained by rabbits from the fat content of whole grains. If you suspect a vitamin K deficiency in some or all of the herd, you can buy K supplements at feed stores handling such supplemental foods, or from the veterinary supply departments of drugstores. Where good clean whole grains are used in the feeding program, there isn't too much likelihood of a shortage of this vitamin.

Vitamin A

This very important nutrient plays more than one role in the rabbitry. It promotes good vision in the herd and helps to ward off infections of the eyes. Vitamin A helps to build and maintain resistance to infection of other parts of the body. It also helps to promote healthy skin and tissue lining and good strong fur.

Vitamin A deficiency can be seen in recurrent eye infections or blindness in young rabbits. It may also be noted in malformed babies. One of the most pronounced of the birth deformities is a bulbous appearance at the front of the head. The little rabbit may appear to have a sac of fluid under the skin of the forehead. This sac feels soft and mushy to the touch. Raising out one of these malformed babies is a real rarity; they usually die within a few days of birth, if they are not stillborn.

Another symptom of vitamin A deficiency is a consistently poor pelt on the adult animal. The skin under the fur may feel dry. In severe cases the skin may be scaly, something like dry dandruff. The rabbit may appear to be in perpetual state of molt (shedding), losing as much or more hair than it reproduces. The coat looks and feels dry and harsh to the touch, instead of having the soft, silky feel of a normal pelt. The healthy coat has a soft sheen to it.

Vitamin A Sources: When pelleted feeds are not used, or used only as a portion of the diet, rabbits can obtain vitamin A from some of the same sources that humans get it. Dark green leafy vegetables and some of the yellow vegetables are good sources.

Rabbits can also obtain vitamin A from good clean alfalfa hay, dandelion greens, plantain leaves, and other edible weeds. It is wise to steer clear of any sort of milkweed for rabbit feeding. The wooly-pod milkweed that grows in the Southwest is the one most often mentioned as being harmful, but I would suggest that even the other types of milkweed be kept out of the rabbits' diet. The wooly-pod milkweed has been proven by scientists to contain a substance which poisons rabbits. The first noticeable results of milkweed poisoning is partial paralysis, evolving gradually into total paralysis and eventual death.

Vitamin D

As long as rabbits are given whole, rolled, or crushed grains to eat, and are raised in outdoor hutches where they receive the benefit of *indirect* rays of the sun, they are not likely to suffer a vitamin D deficiency.

First symptoms of a shortage of vitamin D are likely to be small rabbits having weak, trembly legs. The problem is called rickets, just as it is termed in human children. The little rabbit may have trouble keeping its legs under it, having a tendency to weave, tremble, or even fall over as it tries to walk. It may also have trouble holding its head erect and steady; the head may bobble around or weave to and fro.

If the herd is housed in a dark building, it may be necessary to provide extra vitamin D. This is usually given in the form of a water-soluble powder added to the drinking water. Vitamin D can be obtained at feed stores or veterinary supply departments of drugstores. However, the breeder who gives his herd good clean grains in the feed rations should never have to resort to these supplements. This is doubly true when the rabbits are raised in light airy quarters where the indirect sunlight reaches them.

The B Vitamins

These vitamins are just as essential to the health of the rabbit herd as they are to the human health picture.

Thiamine (vitamin B_1) for instance, helps to maintain normal digestive processes. It also plays a part in keeping the nervous system in a stable, healthy condition. For rabbits, vitamin B_1 is provided in the whole grains of the normal ration.

Riboflavin (vitamin B_2) along with vitamin A, is instrumental in keeping the eyes healthy. It also promotes healthy skin and helps maintain the tissues of the mouth in good sound condition. Vitamin B_2 is provided in the rabbits' diet when whole grains are used in the feeding program. Those green leafy vegetables are another good source of this nutritive element.

Vitamin B_6 assists the rabbit's system in digesting and making full use of the proteins furnished in the diet. Without B_6, at least a part of the proteins the rabbit ingests may not be utilized to the fullest, resulting in eventual protein hunger. As with other nutrients, whole grains such as wheat, oats, barley, milo, the sorghums, and a little corn are basics in providing vitamin B_6. Also potatoes contain vitamin B_6. Instead of tossing the clean potato peels out when potatoes are cooked for the family, save them; while they're still fresh and crisp, serve a piece of raw peel to each rabbit. Of course, any spoiled spots should be cut out before the peels are given to rabbits. Some raisers in other countries concoct a sort of mash made of cooked, cooled potatoes mixed with grains.

Vitamin B_{12} helps to assure proper function in all the cells of the body. Those clean whole grains again come to the rescue.

Vitamin C

Though vitamin C has not been determined necessary to a rabbit's diet, this nutrient does the rabbit no harm. In fact, many people advise giving breeding rabbits a small piece of fresh fruit or fruit peel about twice a week. Most rabbits, especially does and their young, seem to thoroughly enjoy a piece of fruit now and then, though some won't touch it. Citrus peels are a real treat to many young rabbits. Some nursing does appear to appreciate a piece of grapefruit peel occasionally, or even part of a grapefruit, pulp and all.

Calcium

Calcium, in the rabbit diet just as in the human diet, is needed to build strong, healthy bones and teeth. It also helps to promote healthy nerves. Proper function of the muscles and heart is assisted by calcium intake. Given to a doe that is somewhat run-down after kindling and beginning to milk heavily, calcium can sometimes mean the difference between a healthy, well-fed, live litter and a lost one. The nursing doe is most commonly the one that shows a lack of calcium in the diet. A major symptom can be a lack of interest in eating or even a complete loss of appetite.

Adding a teaspoonful of livestock calcium (check with a feed dealer for brand names) to the doe's drinking water each day for about a week after kindling helps to keep the doe in top nursing condition. If she is receiving daily rations of green leafy vegetables, chances are that she is getting sufficient calcium. If the breeder has a surplus of milk (goat or cow), the newly kindled doe may be given a dish of fresh milk each day. Thus increasing a doe's calcium intake for her own system's use and also helps increase a light milker's supply of milk for the baby rabbits. Some raisers soak pieces of very dry, hard, *mold-free* bread in warm milk for their does for a few days after they deliver young, especially during cold winter weather.

Iron

Iron helps to build red blood cells in animals as well as in humans. Once again, green leafy vegetables and whole grains should take care of this need in a herd that is worked moderately. Should you decide to use one of the faster commercial breeding programs and market frying-sized rabbits, you may find it necessary to provide the breeding stock with small amounts of supplemental iron. The advice of a good veterinarian should be sought before adding extra iron to the rabbits' ration, as heavy dosages of iron supplements can affect the liver and its function.

Salt

The use of salt for animals, rabbits included, has come under scrutiny in recent years. While salt may not be an essential nutrient, it does play a very important part in the general health of the rabbit herd. Conversely, lack of salt has been blamed for certain reproductive problems, as well as other ailments. If pelleted feeds are used, the need for additional salt is generally minimal; often enough salt is integrated in these feeds.

When natural grains and other unprocessed foods are used, salt must be provided; the lack of salt in them can cause several problems. Water consumption plays an extremely important role in the efficiency with which rabbits utilize food. With no salt at all in the feed, water consumption is diminished; with water intake decreased, desire for solid food is lessened; with less solid food ingestion, the milking abilities of does are greatly reduced; with reduced milking abilities in the does, the young are denied their full needs for plentiful nourishment in the early days of their lives. Thus, young rabbits may take as long as ten to fourteen weeks to reach the expected normal growth for use as fryers.

So, it stands to reason that rabbits need a small amount of salt along with the vitamins, minerals, and water the raiser thoughtfully provides. Animals given the opportunity will regulate their own salt intake. Food should never be soaked in or covered with salt. Salt should be continuously on the side, so that the rabbits can have it free-choice whenever they feel the need. Some rabbits rarely touch the salt, others eat a great deal. It takes only a short time for the observant raiser to notice which rabbits have the greatest need for extra salt. As a rule, does milking the heaviest require the most salt; they may consume a large quantity when offered a salt spool or lump. These does can be recognized usually by their babies which are fat, full-bellied, fast-growing little animals. And these babies usually begin getting their own little fur coats in a hurry, often within just days after birth.

So, like the combination of vitamin B6 and proteins, salt assists in the consumption and digestion of *other nutrients which are absolutely essential* in the production of ample milk for the young, and in the reproduction of healthy, fast-growing young rabbits for the table.

You can provide salt in the form of the Bunny Spools® that can be purchased in feed stores. Or use hard lumps of ordinary, uniodized canning and table salt. You can even use chips broken from livestock salt blocks when you have other, larger livestock for which salt blocks must be furnished. Some rabbits like the plain white salt while others seem to prefer the sulfur or mineralized salt. Either type can be used.

Since rabbits display a natural wont to gnaw on wood and such, some raisers provide their rabbits both salt and wood at the same time. A piece of two-by-four is soaked in strong saltwater for several days, until the wood is permeated with the salt solution. The block of wood is then removed and dried in the fresh air before it is given to the rabbits to chew.

The whole idea of rabbit nutrition is to provide various foods with which the rabbits can balance their own diets.

Most animals will tend to balance their own nutritional needs if the breeder provides them with the necessary ingredients.

Grains

It is impossible to maintain a healthy, thriftily growing herd on just one grain. Corn alone, for instance, is completely unsatisfactory. But a little corn along with other grains that are higher in nutrients helps to maintain a healthy herd. Neither would wheat or oats alone do a very good job of keeping the herd in the best condition for steady, reliable production and growth.

Greens

Green foods, while they provide essential nutrients for the herd kept on a pellet-free or partially pellet-free diet, are by no means an adequate diet without the addition of grains.

Even the highly praised miracle weed, comfrey, is incapable of maintaining a herd's health and vigor by itself. Comfrey and other green feeds, when properly used, play a very important part in the sound health of a naturally fed herd. But they are simply not sufficient in themselves. No one of these things should ever be relied on to completely nourish the herd. We wouldn't expect a human to stay healthy and vigorous for long on a total diet of one item. Neither can animals!

Remember that whenever green foods are given to rabbits, provide no more than the rabbits will clean up in an hour or less. Stale, wilted greens rapidly lose their nutritional value. Partially spoiled greens can also cause some pretty nasty digestive upsets, even in mature breeding stock. Ingestion of partially spoiled greens can even kill a very young rabbit. Therefore, the removal of any leftover greens in the cage is absolutely necessary! If they're not removed, the rabbits will sooner or later nibble them. The young rabbit with its enormous curiosity will most likely be the very first to eat enough of the spoiled greens to make it sick or kill it.

Another important point to remember when considering nutrition is this: no amount of good nutrition will protect or rescue a herd's health when the rabbits are housed in filthy, smelly quarters! The better the nutrition provided, the longer the strong herd may hold out against these factors. But eventually, poor management will override the benefits of the finest nutritional program, and the rabbits start getting sick.

Feeding the Mature Breeding Rabbit

Most new rabbit raisers tend to overfeed their breeding stock. This is as true of those contemplating setting up a commercial rabbitry as it is of those whose children have acquired Easter rabbits.

To stay in good health and top working condition, mature breeding rabbits from 5½ months of age on, need much less feed than most people think. Working bucks or does without young to nurse need only four to six ounces each of pelleted feed per day. E. J. Duncan of Orange Park, Florida, gives his adult rabbits just three ounces of pelleted feeds per day. To this he adds either two or three large vegetable leaves or a carrot with top attached, as a fresh supplemental source of vitamins and minerals. He also feeds his working stock small amounts of hay to provide additional roughage.

It should be remembered when using pelleted feeds that practically all processors use at least some synthetic vitamins to fortify feeds. Several years ago, one of the most famous commercial rabbit breeders in this country became dissatisfied with the results he was getting from his herd, using one of the best-known brands of rabbit feed. The man had a special finished-product analysis run on a sample of the feed he used. The laboratory which did the analysis discovered that while all of the essential nutrients were present in the mixed but unpelleted feed, there was not even a trace of vitamin E in it.

Some of the nutritional value of feeds is lost in the heat and pressure of pelletizing them. Natural vitamin compounds hold up better during this processing. However, since natural vitamins are more expensive than synthetic ones, the more economical synthetics are chosen for much of the nutritional additive in rabbit feeds.

Whether the adult breeding rabbit is given a diet of all-pelleted feeds, or part pellets with natural foods as supplements, or altogether-home-raised grains and other foods, the breeder must avoid overfeeding. Even a small amount of excess food can rapidly add up to extra weight on a rabbit. This is especially true of the pregnant doe.

A nonpregnant doe should have good firm flesh *except* in the belly; a doe in perfect working condition has a sort of flabby feel to the flesh and skin of her belly. You should be able to gather a fair handful of loose skin on the belly of a doe that has had two or more litters. If you find a doe to have a firm taut belly before she is due to be bred, closely ration her for several days to reduce her weight to the point where the belly once again feels flabby, a little loose skinned. Otherwise, the doe will be entirely too fat when the time comes for her young to be born.

Internal belly fat in a doe can cause an alteration in her metabolism (called ketosis or pregnancy disease)—in which case the doe may die during kindling. Even if the fat doe herself survives the stress of kindling, her young may be born dead or die soon after birth because of the greater-than-normal stress and duration of birth. A doe that is even a bit on the lean side is a better pregnancy risk than a too-fat doe. Another factor in favor of keeping slim and trim is that the fat doe that kindles a live litter continues to convert much of her heavier nursing ration to the upkeep of her own prime, fattened condition, rather than the manufacture of milk for her young.

When the doe is nursing young, it is wise to give her all the food she will eat. Only rarely do you come across a doe that gains weight on a full feeding program while nursing. In our years of rabbit raising, we discovered just one doe that not only fed her young extremely well, but gained weight while feeding them if we didn't watch her food intake closely. Apparently she was one of those does whose system absorbed and utilized every last trace of nutrients in the feed. We didn't dare give her oat hay every day as the other does received when nursing heavily, because she would become too fat in short order!

The buck that works frequently is another matter. He rarely has much of a weight problem, unless he is greatly overfed with fattening grains like oats and corn. In best working condition the buck should be well fleshed, but not fat. Unlike the doe, the buck's belly will be taut, with good firm flesh.

Allowed to become too fat, the buck becomes lazy, not showing much if any interest in breeding. Should he show interest, he is likely to be quickly tired, out of breath, and unable to complete a mating. In some cases of severe overweight, the buck can even become sterile. Fertility may or may not return when the buck's weight is reduced by special rationed feeding.

Both does and bucks are normally expected to have a productive life of from 2½ to 3½ years. However, does have been known to produce well until they were around 5 or 6 years of age. Occasionally, one may find a buck which can sire good litters to the age of 5 years, and more rarely, to the age of 6, if he is worked only once a week. More often, rabbits of these ages will fall off in production. An old doe may drop to only three or four babies to the litter, but still may have an excellent milk supply. She may still be able to raise out nice litters of fryers when she adopts excess numbers from other does' litters. Old bucks may start causing does to miss (fail to conceive), being sterile themselves. There is no known method of restoring productivity to the doe or buck that has passed its prime and gone sterile.

Feeding the Growing Brood Doe or Herd Buck

Many problems with a first-litter doe having an abortion or misconception stem from feeding practices while the young doe is growing to maturity. For the first 4½ to 5 months, the doe should be given full feed (all she will clean up in a twenty-four-hour period). By the time the doe is 5½ months of age, she should be cut back to the same amount given to fully mature does.

If the young doe appears to be getting too fat, it may be necessary to cut her feed even a bit more, to the point that loss of weight becomes barely discernible. This will be noticed in the less taut belly. The young doe will have fairly taut belly muscles until she has kindled and nursed a litter or two, but you can soon learn by the feel whether it is muscle tautness or fatiness.

Young bucks should be full-fed until they reach 5½ to 6 months of age. At this age, their rations should be the same as those of mature working bucks. Each rabbit requires a slightly different amount of feed to hold its weight at the best level for its own ideal working condition. The individual raiser thus has to work out his own feeding program, taking notice often of the general condition of each rabbit.

Once young rabbits start working regularly, they should be treated as fully adult, working animals. Does need just as much nutritious feed to build milk for their first or second litters as older does need to maintain their eighth or tenth litters.

Feeding the Weaned Fryer

While the young rabbits remain with the mother doe, they should receive full feeding. They still rely on the mother's milk to some extent, also. Once the young are weaned away from the doe, they should continue to receive full feeding— that is, as much feed as they will clean up in twenty-four hours. Fryers should never be allowed to go without feed for hours at a time.

The first and most obvious result of fryers not getting all they can eat is noted in their flesh conditions and growth rates. Bone structures of young rabbits grow rapidly, which calls for plenty of feed to produce ample flesh to keep up with the fast bone growth. Fryers being underfed will quickly look skinny and poorly furred.

A second result of underfeeding fryers is seen in fur eating. With too little in their bellies to meet the demands of growing bodies, the little rabbits turn to the next most convenient thing—trying to appease hunger by eating the fur of their littermates and themselves. Nutritional hunger (lack of essential nutrients in the feed) causes fur chewing, too. Protein starvation causes hair chewing among does and bucks as well as fryers.

Very young rabbits must be rationed carefully when provided with green feeds such as garden vegetables and edible weeds. When they start coming out of their nest, the amount of greens should not be increased right at first to provide extra for the young; just give the doe her usual ration, and the young will get a few nibbles of that. Gradually increase the amount of green foods until the young are about six weeks old. Never give more than what the rabbits will clean up within an hour, as stale greens can cause diarrhea in the young. By gradually introducing green feeds to the young rabbits' digestive systems, you can eliminate much of the digestive upset which has caused some raisers to discontinue the use of any greens at all.

Oat straw is an excellent source of roughage for the young rabbit. On the other hand, oat hay with a great deal of grain left on is not very good. First, the rabbit tends to gorge itself on the raw grain and become sick. Second, the husks of the oat grains may cause damage to the stomach and intestinal tract of the very young rabbit. If homegrown, natural grains are fed to young rabbits, first crush them and remove most of the husks. Rolled oat groats are an excellent beginner feed for the little rabbit just starting to eat.

If concentrated protein supplement pellets (such as Calf Manna®) are used for nursing does to promote milk production, care should be taken that the small rabbits do not devour too much of these. Too much protein in a concentrated form can cause diarrhea in very young rabbits. For the fryer weaned away from the doe, you may use small rations of protein supplement, but do not give the litter enough of the supplement that any one of the fryers can gorge itself on it.

Feed Formula

In recent years, quite a number of rabbit breeders have been dissatisfied with the results they've obtained from previously fine herds, and many have had analyses done on their usual brands of feed. In several instances due to production costs, ingredients in the feeds had been altered; tests showed that fat content was higher than required by producing rabbits, while protein and vitamin content had decreased.

The breeder mentioned earlier, who had one of the first finished product analyses run on his brand of feed, studied rabbits' nutritional needs and also nutritional values of grains and so on, and then he worked out his own feed formula. The amounts of ingredients in this formula are originally meant for the large herd, but they can be used equally well by the small raiser if he has access to a feed mill which will custom-mix feeds. Most feed mills can obtain the necessary ingredients, and a competent miller can adjust the various measure for smaller batches of feed than the ton and one-half of pellets made from the following formula.

Feed Formula:

700	lb.	ground barley (only heavy, clean A1 quality; 48 lb. or more per bushel). In areas where barley is not readily available, cornmeal may be substituted.
170	lb.	ground wheat (A1 quality, cleaned).
200	lb.	ground oats (must be 32 lb. per bushel or better, A1 quality, cleaned and ground finely).
100	lb.	mill-run wheat bran.
1,400	lb.	alfalfa meal (½ sun cured), ½ dehydrated; A1 quality).
25	lb.	distillers' dried grains with solubles (corn).
370	lb.	soybean meal (48½ percent protein).
100	lb.	brewers' dried grains.
30	lb.	granular bentonite.
33	lb.	phosphate (21 percent grade).
20	lb.	salt.
75	lb.	lignin sulfonate.
25	lb.	animal fat (optional).
50	lb.	premix*.
25 to 30	lb.	molasses has been used per batch.

*Premix: (50 pounds)

¾	lb.	trace mineral.
2½	lb.	choline chloride.
$1/16$	lb.	vitamin E 100,000 concentrate (extra).
3	lb.	mixed vitamins of the following potencies per pound:

3,000,000 units A.

833,333 units D_3.

5 mg. B_{12}.

3 g. riboflavin.

3,000 units E.

2 g. medadione sodium bisulfite complex.

17.5 g. niacin.

250 mg. folic acid.

4.5 g. d-pantothenic acid.

.5 g. pyridoxine hydrochloride (vitamin B_6).

.5 g. thiamine (vitamin B_1).

The preceding vitamins are multiplied by three to get the total amount added per batch of feed. Ground milo is added to the premix to make fifty pounds. Twenty grams of Ethoxquin per pound of vitamin mix may be used to preserve freshness. If hair chewing is a real problem in the herd, one pound of magnesium oxide can be added to the premix. This acts as a preventive. As a rule though, the well-nourished herd does not require any hair-chewing preventive. Most hair eating, in our own experience, can be corrected by supplying a better grade of protein in the feed.

Natural Feeds

Many people make the mistake of believing that rabbits can do well on just one grain with a few garden greens added for variety. Corn is the usual choice, since it normally is less expensive than other grains. As anyone who has ever had to resort to a reducing diet can tell you, corn is fattening! It has its place in animal feeding, but production rabbits can rapidly become too fat on a diet of corn alone. Such a limited diet can also cause protein and vitamin hunger, leading to other health problems.

Here are the grains, along with their protein contents, that are preferred by successful rabbit breeders: corn, approximately 10 to 11 percent protein; oats, approximately 12 percent; barley, approximately 12 percent; buckwheat, approximately 10 percent; and wheat (soft), approximately 11 to 13 percent. To provide altogether natural feeds, you can mix the following for dry does, young rabbits, or herd bucks: equal parts of crushed wheat (or oats or corn) and barley, milo, or other grain sorghums.

Usually considered only as a source of supplemental protein, soybean meal contains approximately 45 to 48 percent protein. It may be added to the rabbit ration at a percentage of 5 to 7 percent of the total ration. Adding too much soybean meal can cause digestive problems in all rabbits, but especially in the younger stock.

Peanut meal contains approximately 47 percent protein; therefore it should be used with caution to prevent the same problems which soybean meal can cause.

Cottonseed meal contains about 41 percent protein, but is not recommended for rabbit feeding unless one can obtain what is called degossypolized meal. Cottonseeds contain an ingredient called gossypol, which is not digestible by one-stomached animals such as rabbits. If you can find the degossypolized cottonseed meal, you can use it sparingly as a supplemental source of protein.

Hays useful in feeding the rabbit herd are: prebloom alfalfa, approximately 19 percent protein content; early bloom alfalfa, about 16 percent; common vetch, about 17 percent; annual lespedeza, about 15 percent; bluegrass hay, about 12 percent; and red clover hay, about 13 percent. Protein content of alfalfa deteriorates as the plants grow older. If carbonaceous hays are used in feeding rabbits, it is advisable to provide some extra source of protein. Grasses such as prairie, timothy, Johnson, Sudan, Rhodes, and others contain less protein than the legume hays such as alfalfa, clover, vetch, kudzu, lespedeza, and peanut.

Feed the legume hays free choice in a hay manger or rack on the side of the cage. All hays (and other feeds) should be completely free of molds; molds can kill rabbits, as well as larger animals.

Oat hay is not a good source of protein, containing only about 4 percent when cut in early bloom. Peanut and soybean hay contain approximately 5 percent protein. Soybean hay is not recommended as it is believed to be rather unpalatable to rabbits, quite indigestible, and a possible cause of reproductive problems. Full-grained oat hay, while not a good source of protein, is very palatable to heavily nursing does and apparently helps them to produce more milk for their young.

In addition to the grains and hays that you may use to feed the herd, you can also make use of many of the vegetables grown in the garden. A comfrey leaf once a day, a collard leaf, or a carrot with the top left on give additional vitamins, plus some extra protein. Small amounts of additional protein won't harm the healthy rabbit, but greatly overfeeding with protein may cause loose bowels, especially in younger stock. Carrots are a good source of carotene for rabbits. The carrot tops should be fed along with the washed carrot roots. Dark green, leafy vegetables are good sources of vitamins—vitamin A, one of the major ones. Edible weeds such as plantains and dandelions are excellent sources of vitamin A.

You may provide potatoes or their peels as a supplementary food, as discussed on page 90. Some rabbits like turnips, sweet potatoes, parsnips, and the like.

Various types of pea vines from the garden may be used fresh or made into hay. Rabbits love fresh English pea vines. Green bean vines and the vines of other beans (excluding the soybean plant) may also be fed to rabbits, either green or dried for winter feeding.

Comfrey is a highly recommended source of protein for rabbits. If comfrey is fed regularly, either fresh from the garden or in the form of hay, there is no need to provide additional protein. This is also true when clean, sun-dried alfalfa hay is used. Given a choice, rabbits will invariably pick sun-dried alfalfa over the dehydrated kind. Dehydration may destroy many of the vitamins in alfalfa.

In addition to a high protein content, comfrey contains a substance called alantoin, used in medicinal products. It appears to help keep rabbits their healthiest, and it is frequently recommended by natural food enthusiasts as a means to prevent or cure the diarrhea of young rabbits.

Wheat bran contains approximately 16 percent protein plus vitamin E. Wheat germ may be used as a vitamin E source, or five or six drops of wheat germ oil on top of the dry feeds. Do this especially if pelleted feeds are used exclusively.

Very dry bread may be fed to rabbits if it is completely free of mold. Some raisers use dry bread soaked in surplus goat or cow milk as a supplement for young growing rabbits, and sometimes this is also used for heavily nursing does.

For the doe with nursing young, it may be necessary to give protein supplement. This may be done by feeding the doe two or three extra comfrey leaves. She should also have good clean legume hay, fed free choice in a rack. In the abscence of the legume hay, extra protein may be provided in small rations of soybean meal cake or pellets, peanut meal cake, or degossypolized cottonseed meal pellets or cake. Caution should be used in feeding these concentrated types of protein supplement, however, since they can cause digestive upsets even in mature rabbits. One tablespoonful of these proteins is usually sufficient for a milking doe per day.

We always found it helpful in assuring a doe's good milk supply to give each newly kindled doe a few stalks of clean, *unsprayed* oat hay as soon as possible after her young were born. Each day thereafter for the next three weeks (the heaviest milking period), each doe received her ration of oat hay. Each doe also received all she would eat of the regular rabbit feed; since we used mostly pelleted feeds, we also added a tablespoonful of protein concentrate. Never skimp on feed if you expect to have nice, meaty fryers in a reasonable length of time. A heavily nursing doe will usually eat around twelve to sixteen ounces of feed per day plus her protein supplement and the hay. If she runs out of feed long before the next feeding, her milk supply for the young will deteriorate. A nursing doe will drink around a quart of water in twenty-four hours.

A group of eight or nine, seven- to eight-week-old fryers, weaned from the mother, will normally eat about six ounces of feed per rabbit per day. In addition to this, they should have whatever fresh foods they can clean up within an hour, plus extra protein in the form of free-choice legume hay or comfrey leaves. Oat straw is an excellent form of roughage for growing rabbits, and it won't cause digestive problems since it has a low protein content. In fact, it has been known to stop certain types of scours in young rabbits.

Dry does and working bucks normally require about five to six ounces of pelleted feed or unpelleted grain mix.

By supplementing with fresh, clean garden foods, you can cut this amount to some extent, but give each rabbit at least three ounces of pellets or mixed grain. These measures are for the medium-size breeds, of course. The tiny breeds of rabbits do well on approximately half the amount of feed given the medium breeds.

Colonizing a Herd

To cut down feed costs, some people advocate putting rabbits into a fenced pen on the ground, so that they can eat the grasses and weeds there. This is called colonizing or raising in colonies.

Colonizing is not advisable; transference of diseases is the reason. Rabbits are not particularly fastidious eaters. Raised in soiled quarters, or on the ground in large groups, they tend to eat the droppings of other rabbits. This can transfer intestinal parasites from older rabbits to the young. Coccidiosis is one of the major problems transmitted in this manner. Cocciciosis damages the livers of rabbits, making them inedible. It is considered one of the most important disease factors in losses of young rabbits in the United States.

Breeding cannot be controlled in the colonized rabbit herd. A buck rabbit will mate with any doe that is handy. It makes no difference to him whether the doe is his mother, sister or whatever. Inbreeding can be useful when properly controlled, but it can rapidly downgrade the herd if not controlled.

Colony raising can also literally destroy any control the breeder would have over the spread of diseases. Rabbits are sociable animals—touching noses, eating together, playing together. In a colony, one sick rabbit can easily infect every other rabbit before the raiser even realizes it is ill.

While raising rabbits in colonies on the ground is not a good practice, sometimes it is advisable to put a rabbit that is ill with a digestive upset on the ground in a protected wire enclosure. It has been scientifically proven that clean soil contains certain antibiotics, microbes, and such, that are helpful in correcting scours in young animals. The additional exercise the rabbit gets while exploring its quarters on the ground is also helpful in curing digestive problems.

Chapter 5

Ailments

When well fed and properly housed in draft-free, sanitary hutches, rabbits will be relatively free of diseases. The best cure is always prevention.

But rabbits do get sick, just like people, and just like other animals. Unfortunately, not a whole lot is known about all the diseases to which rabbits are prey, simply because very little research is done on rabbit husbandry. In part, economics contributes to this problem for the rabbit raiser. It is ofttimes less costly to dispose of a sick rabbit than it is to take it to a vet, who may not know a whole lot about rabbits, and in addition pay for medication. Many regard disposal as undeniably surer, since the sick rabbit's ailment can't infect the rest of the herd if you quickly dispose of the sick rabbit.

In the following pages, I will explain what I know about rabbit ailments, together with suggestions on preventing and curing these ailments.

Natural Animal Medicines and Their Use in the Rabbitry

Since the advent of chemical medications in the field of animal husbandry, many people scorn the use of natural medicines as old-fogey, hopelessly out of date. The truth of the matter is that acquiring these natural medications with which animals in the wild treat themselves is usually more work, more of a problem than is met by dosing the animals out of a bottle or a packet of ready-made medicine.

Turn a sick little rabbit loose in an enclosed pen on the ground where plantain or dandelion grows profusely, and watch it attack those weed leaves. It sometimes appears that the little animal is gobbling so much of the leaves that surely the stuff will kill it! Yet, wait a few hours, and see how the nastiest case of diarrhea will be cleared up. As is the case with balancing its own nutritional needs when it can come by the required ingredients, that little rabbit somehow has an innate sense of what is needed to make it well.

A great many old-time rabbit breeders state without hesitation that plantain weed leaves are one of the best preventative medications the raiser can furnish his herd. A close second is the leaves of the elm tree when these are available. These two natural materials are excellent for either curing or preventing the common scours (diarrhea) that sometimes afflict the very young rabbit.

Feeding the leaves of plantain, dandelion, or the leaves and some bark from elm once or twice a week can help keep a rabbit herd from developing any need for the more drastic (and sometimes less reliable) chemical drugs. The only medication I ever used in our small commercial herd, that came anywhere close to giving the same good results as the natural medicines, was not a medication in the real sense; it was a combination of vitamins which is no longer on the market because it had been developed for chickens but it gave no immediately visible results in poultry herds.

Common sassafras, used sparingly due to the possibly harmful saffrole content, has a beneficial effect on nursing does and young rabbits old enough to eat some solid foods. A piece of sassafras limb with the bark left intact, given to a doe and litter once every two or three weeks, also helps to keep the babies free of digestive upsets. Once the bark is all eaten from the piece of sassafras, it is not uncommon at all to see that the doe has dunked the wood into her water bowl, if a crock is used in the cage. The wood stains the drinking water a pale yellowish color, similar to a very weak sassafras tea. She and her young will drink this tea with great relish while it's fresh. Remove the wood from the bowl and refill the bowl with clear water, and the doe will pick the wood up in her teeth and, to all appearances, impatiently dunk the wood right back into the bowl of fresh water. It's just as if the doe is telling the breeder that she knows what is best for her brood!

Blackberry leaves have long been considered an excellent cure for scours by longtime breeders. One thing is certain—rabbits dearly love blackberry leaves. Don't ever attempt to tease a special pet rabbit with one of these leaves; you may wind up with a badly bitten finger or two!

Luckily, the dried leaves of certain trees are also enjoyed by rabbits, and they apparently have close to the same effect on their health that the green leaves have. The dry leaves should be gathered when they are fresh and clean, free of mold, decay, and insects. Rabbits seem to enjoy best the dried leaves of oak trees, elm, sassafras, and willow.

Pieces of willow branches are also regarded highly by many who raise rabbits without the use of chemicals other than a coccidiostat. When feeding small pieces of tree limbs, always leave the bark on. Apparently the bark is what contains substances that are beneficial to rabbits. As with garden produce, make sure that any leaves remaining after an hour are removed from the cage. The dry leaves may be fed in the same manner as hay—in a rack fastened to the door or side of the hutch. Without the high moisture of the green leaves, the dried ones do not cause digestive upsets in even the very youngest rabbits.

Grass clippings may be given to rabbits, provided the grass has never been sprayed with insecticides or herbicides. One example of a rabbit subjected to these chemicals is a doe belonging to a breeder in Florida. This doe somehow managed to escape her cage. While she stayed in the breeder's own yard she was safe, but she wandered through the fence into a neighboring yard where insecticides were used. The breeder caught the wandering doe and returned her to the cage, but he was too late. A few days later the doe lost her sight completely. To add to the breeder's distress, the doe was pregnant and went on to kindle her litter in the nest box by instinct alone, since she could no longer see the box. Shortly after delivering her babies, the doe died from chemical poisoning. Her litter was also lost, of course. A rabbit raiser thus should be very cautious in feeding grass cuttings to his animals. Grass or hay that is cut along a highway or county road should not be used for rabbits, either, since there is a possibility of a too-high lead content.

Apple cider vinegar is at last coming into its own in many rabbitries across the country, as well as in Canada. When I was rabbit editor for *Countryside and Small Stock Journal*, I received a number of letters from other rabbit raisers espousing the benefits of using cider vinegar.

Since apple cider contains good amounts of potassium, perhaps this is the major benefactor for rabbits. Vinegar diluted to 5 percent acidity is recommended; one or two tablespoons are added to a gallon of fresh, clean drinking water.

Many raisers with diarrhea trouble in their young stock claimed that after three months of regular vinegar dosage, the scours ceased to exist in their herds. Problems with scours don't disappear overnight, of course; the vinegar must be continued on a daily basis to obtain the best results. If vinegar-treated water is not given daily, the effects are likely to be spotty and unreliable for controlling any sort of disease problem. Some breeders claim that giving rabbits this water-vinegar daily even keeps those troublesome coccidiosis germs under total control.

Another benefit of the apple cider vinegar has been reported. Breeders who had experienced trouble in getting their does bred have stated that after giving them cider vinegar in drinking water for periods of up to three months, their does became not only willing, but actually eager to mate with bucks, and they went on to deliver fine healthy litters.

Apple cider vinegar also appears to improve the rabbits' coats, giving them a healthy sheen even in hot months when rabbits are likely to be shedding at least some of their heavy fur. Improvement in rabbits' coats is probably due to the acidity of the vinegar, which can affect the pH balance of the skin and give the hair follicles a healthier base in which to grow.

Comfrey, used properly, can be one of the most useful plants for the raiser who wants to produce rabbits as naturally as possible, since it contains most of the amino acids that are the building blocks of protein, plus the curative substance allantoin that is used in numerous medications. A great many of the most common rabbitry problems can be traced to a lack of protein. Comfrey fed on a regular basis combats these protein deficiencies.

The real beauty of comfrey in the rabbitry is that it can be used as a fresh green during growing months, or dried and fed as hay during winter months. Like alfalfa, it should be fed as fresh as possible to have the greatest effect, but drying the summer crop for winter use is as practical as using other hays while it offers more extensive benefits with regular use. Benefits from comfrey have never been conclusively proven by scientists, but this plant has proven extremely helpful in numbers of private rabbit herds.

Rabbit breeders who have used comfrey report that they rarely, if ever, have rabbits sick with colds; that comfrey has saved numerous young rabbits suffering from various types of diarrhea; that does suffering run-down conditions from heavy work schedules and nursing have shown excellent recoveries when given comfrey to eat regularly.

In raising rabbits, you can come in contact with some of the strangest methods of treating diseases. Pekoe tea given in drinking water is one of these odd-sounding remedies, but it has been recommended by an animal research veterinarian at a famous feed company's own laboratories in Missouri. When various antibiotics failed to cure a killer disease in rabbits known as Tyzzer's disease, the veterinarian suggested the use of black pekoe tea diluted at a ration of one large tea bag to one gallon of water, cooled and given to the sick rabbits as drinking water. The tannic acid in the tea kills the germ that causes Tyzzer's disease. With further experimentation by raisers, the weak tea also helped clear up cases of ordinary diarrhea when given with clean, bright straw to eat.

As with other useful remedies, the tea should be given only when a real need is determined. Otherwise, the animals may build an immunity to beneficial ingredients in the medication.

Stress and Stress-Related Diseases

First of all, it is important to emphasize that a great many of the health problems experienced by domestic rabbits originate in stress. Which leads to the question: Just what is stress, and where does it come from?

The usual definition of stress in connection with cage-raised rabbits is *any deviation from the normal sequence of events in or around the rabbitry*. This can be strange people wandering around in the building; it may be unusual sounds in or near the rabbitry. Stress can be caused in rabbits by a dog or cat prowling around where the rabbits can smell or hear it.

A severe lightning and thunderstorm will sometimes stress the animals, pregnant does in particular. Nutritional deficiencies, either sudden or long-standing, can create a severe stress situation, especially among pregnant or nursing does and young rabbits. Stress is not normally so quickly detected among mature dry does and herd bucks as it is among pregnant or lactating does and young bunnies. Both lactation and pregnancy put a good doe under at least some amount of stress. A very young rabbit grows so rapidly that the growth itself can be a stress factor. Any additional stress may be just the factor that makes the rabbit sick.

Extreme heat during the summer is one of the most common stressful situations that affect rabbits. Many people are surprised—some actually disbelieve—that caged rabbits can suffer heat prostration just as humans can.

While pregnant does, as previously mentioned, are most susceptible to heat stroke, all rabbits suffer from extremes of heat.

Cold is another factor which can cause stress, but not nearly as often as heat. As a rule, stress from cold shows up in sterility among mature adults. This usually doesn't happen unless the temperature drops to an extreme low and stays there for some time. As long as the rabbits are protected from cold drafts to prevent colds and pneumonia, only prolonged extreme cold should cause trouble in the rabbitry.

Noises may be a problem for those raising rabbits in or near a city or an airport, and probably those unlucky enough to live directly under a major airline's flight lane. Large machinery working nearby can keep rabbits upset to such an extent that they eventually cease functioning normally. This noise factor may be brought on, too, by children playing rowdily in the immediate vicinity. Normal human speech and activity do not seem to upset most caged rabbits; it is the sudden, unusually loud squeal or scream that does the damage. Dogs barking suddenly can startle them, causing nervous upsets.

Rabbits can and do adjust quite well to constant noises such as cars passing by. But most of these animals are easily startled when they hear a sudden, unexpected loud noise, especially if the source of the noise is quite near to them.

Lack of proper sanitation in the rabbitry probably causes more year-round stress-based problems than any other single factor. Living in filthy quarters can stress any animals, including those much larger and stronger than rabbits. Accumulations of manure and urine create an ammonia gas in the building. This in turn causes a severely oppressive strain on the herd. As with heat, does carrying or nursing young and little rabbits react most rapidly. When the rabbitry operator can smell ammonia gas in the rabbitry, damage is being inflicted in the lungs, nasal passages, and bronchial tubes of the rabbits.

Nutritional Deficiencies: Nutritional stresses are harder to correct than those brought on by weather and noise. Inadequacies in the day-to-day diet of the herd can crop up in some of the strangest ways. One of these problems appears to affect only very young rabbits, though the actual deficiency is experienced by the mother doe. The doe herself won't normally show any outward symptoms, ex-

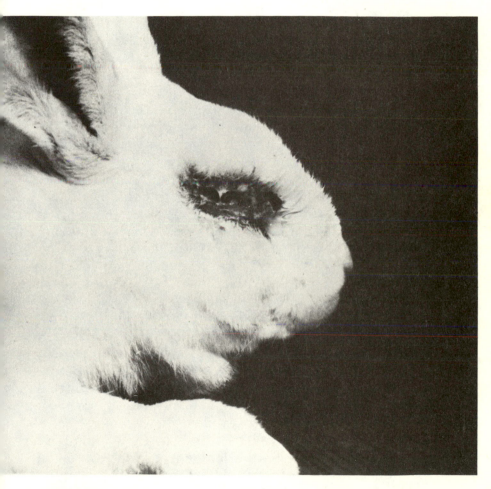

Pasteurella infections can produce a variety of external symptoms, in this case, conjunctivitis or weepy eye.

cept that she may be somewhat thinner than usual. The tiny bunnies, anywhere between a few days and several weeks old but still relying mainly on the doe's milk for nourishment, develop tiny bumps or pus pockets. These tiny pustules most commonly are noticed on the necks and bellies of the little rabbits. When the babies are a bit older, already furred out, the first indication of any problem may show up in white scaly-looking material around their noses (similar in appearance to dry, flaky dandruff). The fur around the noses may shed, leaving only bare but scaly, snowy white patches on either side of the nostrils.

When this problem appears in the young bunnies, the mother doe is carrying a nutrition-deficiency disease germ called *Pasteurella*. This disease is contagious through direct contact between rabbits, so those affected should be isolated. The doe and her litter may be kept together and medicated to correct the problem.

Vitamin therapy on a daily basis is the only cure for this type of *Pasteurella*. Even if the young are still nestlings, not yet eating solid food or drinking water, vitamins given to the doe will benefit the young as well. The vitamins should be administered through the doe's drinking water.

Pasteurella bacteria are responsible for numerous external symptoms of other diseases. For instance, the caked breast which has no other apparent cause can sometimes be traced to an increase of *Pasteurella* organisms in the doe resulting from some form of stress. While *Pasteurella* germs can be intensified in a lactating doe by a shortage of proper nutrients in her daily diet, these microorganisms are more often given an ideal breeding ground when the animals live in dirty quarters, breathe foul air, and are subjected to chilly, damp conditions.

Pseudotuberculosis is another *Pasteurella*-created disease. It results in the development of nodules resembling the tubercules of tuberculosis, which may be noted in the intestinal wall, lungs, spleen, and liver. External symptoms include anorexia (lack of appetite), emaciation, and general lassitude. The disease is contagious, not only to other rabbits, but also to man. Therefore, any suspect animals should be destroyed and buried quickly; the meat is not edible. Hutches or cages should be disinfected with a very strong, hot disinfectant solution promptly.

Pasteurella germs are also behind some of the abortions and miscarriages that occur. Again, the unsanitary, ammonia-fumed building is most frequently the causative factor. Or, when just one doe out of the entire herd repeatedly aborts or miscarries her young, a lack of stamina in that one doe is probable. However, if several does, especially from different family lines, abort or miscarry often, most likely there is a nutritional inadequacy somewhere in the daily ration. Vitamin E is the vitamin considered most important in maintaining the reproductive abilities in animals, so abortions and miscarriages among previously fine does is a signal that *natural* vitamin E, and quite possibly other important nutrients, are not being furnished in sufficient quantities.

Stress-Related Bacterial Diseases: Other bacterial diseases, not caused by the normal stresses as are the *Pasteurella*-based ailments, are not too common in the average herd, but some knowledge of them is important since a couple of them are transmissable to man.

Toxoplasmosis is one of these rarer bacterial diseases. It is a protozoan infection. Believed to be passed from cats to rabbits through fecal contamination, *Toxoplasma gondii* is presently classified by scientists as a coccidium. As such, sanitation in the rabbitry and faithful execution of a coccidiosis prevention program are the raiser's best insurance.

The sexual phase (formation of oocytes or eggs) takes place only in the intestinal tracts of the cat family. Symptoms include anorexia, emaciation, tremors of the nervous system, enteritis (diarrhea), and sometimes a lack of coordination or paralysis. In addition, some abortions or miscarriages have been traced to this disease.

Should you suspect that a rabbit in the herd has somehow contracted toxoplasmosis, a good veterinarian is your best protection; the disease is transmissable to man if proper protective measures are not taken in examining and disposing of the infected animal. No type of home treatment is recommended, whether by chemical or natural medications.

Prevention of any factors which can multiply cocci germs is a must. When cocci bacteria are allowed to infect a herd over a prolonged period, toxoplasmosis can be a dangerous result.

Tularemia, also known as rabbit fever, is rare among rabbits raised in cages, but they are susceptible to it. The disease is usually picked up by a domestic rabbit when it is bitten by a rat which can also carry the germ. Rabbit fever is not carried by domestic, cage-raised rabbits as it is by wild rabbits.

The germ responsible is called *Pasturella tularensis*, and it is passed to animals in the wild by a wood tick, *Dermacentor andersoni*. Any wild rodents, but especially wild rabbits, are affected, as well as sheep in some areas. Humans are also susceptible; exposure most commonly occurs in cleaning infected rabbits or eating the undercooked meat thereof.

Tularemia may occur as only a latent infection without clinical symptoms. If an animal is heavily infected, it may have a fever of 105 to 107 degrees Fahrenheit. (Normal temperature for an adult rabbit is 102 to 104 degrees.) The animal may experience increased respiration, diarrhea, general stiffness of the joints, coughing, weakness and lack of coordination, and general fatigue and depression.

Treatment with the tetracycline drugs may benefit the animal which has been diagnosed as having tularemia. Oxytetracycline given at levels up to 5 milligrams per pound of body weight daily is particularly effective. Any diagnosis or treatment should be done by a qualified veterinarian.

Prevention is maintained by keeping the area around outdoor hutches clear of ticks; by not having any wild animals, especially wild rabbits, in the rabbitry; and by eliminating any rodents such as rats in the buildings or hutches.

Tyzzer's disease is another bacterial disease that sometimes attacks rabbits. It shows some characteristics of bacteria and some of virus in the causative agent, which is the infectious organism known as *Bacillus piliformis*.

The first cases diagnosed were found among Japanese waltzing mice in 1917. In 1965 the disease began showing up in rabbitries in the northern part of the American continent, but it is believed to be frequently misdiagnosed as enteritis, coccidiosis, or in adult rabbits as heart failure.

Tyzzer's can be a swift killer of young animals. Fryers that appear to be well in the morning may be dead by evening. But infected animals may live a couple of days. The primary symptom of Tyzzer's is a persistent, very dark-colored diarrhea. When a postmortem is performed on a dead rabbit, one may find inflammation of the cecum (lower gut) with surface hemorrhages. Death of some tissue in the cecum, upper colon, and lower ileum may be noted; the tissue will appear yellow and wrinkled.

Because the disease affects primarily the digestive organs, it is believed to enter the system orally and to leave the body in the feces. Fecal contamination of feeds, pens, water dishes, equipment, and handlers is considered the most likely means of transmitting the disease from one animal to another. Not much is really known about the causative organism, but it is suspected that the transfer of carrier animals through sales and shows infects a number of healthy rabbits.

When Tyzzer's has been definitely diagnosed in the very small raiser's herd, probably the best solution is to dispose of all rabbits, thoroughly clean and disinfect all nondisposable equipment, burn all straw and hay that has been in the rabbitry with the afflicted animals, and start over with hopefully clean rabbits. The only certain method of determining whether or not an animal is infected is to examine the liver and other digestive organs, which means killing the animals.

Stresses from overcrowding, heat, and other such factors can make it easier for the organisms to gain access to the intestines, from which they travel into the bloodstream and thus to the liver.

Treatment with antibiotics, under the supervision of a good veterinarian, may help to hold the bacteria under some measure of control but will not cure the problem. The meat of an animal which has recovered from Tyzzer's disease is edible; the liver is not.

Mice are believed to be responsible for some herds becoming infected with Tyzzer's. It is advisable to keep mice out of the rabbitry and all feed and equipment storage areas. Hay or straw should be checked for signs of mouse infestations before use in feeding or as nesting material.

Ordinary black tea that is used for human consumption has been prescribed in at least one outbreak of Tyzzer's in a small herd, by a research veterinarian in Missouri. The tea is made by steeping one tea bag in one gallon of hot water, letting it set until cool, then giving the resulting weak tea to affected rabbits to drink. Acid in the tea apparently controls the organism which causes Tyzzer's.

Coccidiosis

Causes: Coccidiosis is undoubtedly one of the most common and most costly disease problems that afflicts the domestic rabbit herd. There are two types of coccidiosis: hepatic (liver) and intestinal. Both are caused by a parasitic infestation of cocci organisms. Liver coccidiosis is caused by an organism known as *Eimeria stiedae*. Intestinal coccidiosis can be caused by infestations of any of four related germs: *E. magna, E. irresidua, E. media,* or *E. perforans*.

Coccidiosis is perpetuated in a herd through ingestion of contaminated matter. Rabbits often pick up and eat droppings of other rabbits or their own. The cocci germs are carried in these droppings, as well as those of other animals such as dogs and cats. For this reason, the all-wire floor in the rabbit cage is recommended, and all rabbit feeds should be kept sealed away from cats and dogs. Hays and grasses for feeding should be kept where dogs and cats cannot sleep on them. While cocci germs seem to be more or less inherent in all herds, heavy infestations from outside the herd appear to be harder to eliminate than the organisms that the rabbits themselves carry.

According to some experts, baby rabbits are born free of coccidiosis, but they usually pick up the germs in short order once they are moving around. This occurs when the little rabbits lick their soiled feet, ingesting droppings in the nest box or on the cage floor, or drink water or eat feed that is contaminated with fecal matter of other rabbits.

Symptoms: In minor infestations there will be few or no outward symptoms. When a rabbit is slaughtered, whitish spots may be discovered on and in the liver when hepatic coccidiosis is present; this liver is inedible, but the flesh is considered edible. The victim of liver coccidiosis suffers diarrhea and anorexia (loss of appetite), fails to make expected weight gains, and usually has a rough coat. In severe infections, a young rabbit usually dies within a fairly short time. Length of illness before death depends on the severity of infection and the basic genetic strength and stamina of the animal in question. Rabbits that recover are not normally immune to further infections.

Intestinal coccidiosis is often confused with other diseases—mucoid enteritis in particular. There is a tendency toward soft droppings. An autopsy won't usually show any lesions unless a microscope is used. Mortality is not too common in a mild case, but when the infection is heavy, there is likely to be profuse diarrhea, weight loss, and death.

Contagion: Coccidiosis is highly contagious within the herd if the germ is allowed to continue untreated; however, the five types of cocci organisms inherent in rabbits are not known to infect humans or other animals or poultry.

Treatment: The only scientifically proven method of treating coccidiosis is a sulfa drug, which is usually given in drinking water. Coccidiostats come under various brand names having sulfa as the basic ingredient. The brand may be Sulmet®, which is a liquid. A water-soluble substance named sulfaquinoxaline is used by many larger breeders, but this type usually comes in packets larger than are practical in the very small homestead rabbitry. Whatever the brand name of the medication, when properly used, each is equally effective.

Package directions should be followed faithfully. If the packet does not give directions for dosing rabbits, you may use the instructions for mink.

Prevention: For routine prevention, the medication is given for three consecutive days out of each thirty-day period. This will keep the *E. stiedae* under control, but it will not prevent infections from the four germs which cause intestinal coccidiosis. A ration containing 0.1 percent sulfaquinoxaline fed continuously for two weeks will help control intestinal cocci.

Sanitation is the surest method of helping to prevent subsequent infections by either type of coccidiosis. Without proper sanitation, no medication is going to keep a rabbit herd free of frequent infestations by any one of the cocci organisms.

Colds (or Snuffles)

Causes: Colds among rabbits are often called snuffles, but some raisers are beginning to feel that there are two different problems here.

The typical cold usually clears up with treatment and, if it is a true cold, doesn't usually return unless the rabbit is further exposed to the drafts or damp living quarters that caused the cold the first time.

This runny nose is a sure sign of snuffles.

Here's how to administer a shot. It takes two: one holds the rabbit by the shoulder skin and the rump, and the second grasps the hind leg and gives the shot.

On the other hand, I think of snuffles as a more or less constant problem an occasional rabbit may have. The apparent cold recurs time after time, regardless of treatment or living conditions. The ailing rabbit may appear to respond to treatment, but as soon as medication is stopped, the sniffly nose returns. The hair on the inside of the front paws becomes caked and matted with exudate from the nose; face fur also becomes matted with the nasal discharge. The rabbit sneezes.

Snuffles may or may not stem from an ordinary cold. More often, I feel—and numbers of other rabbit people agree—snuffles are brought on by other things. Constant irritations of the respiratory system can cause a discharge from the nose, watery eyes, and sneezes and coughs.

These respiratory irritations are usually of a long-lived nature, brought on by some constant factor in the rabbitry. This may be excessive dust particles in feeds, hays, or grasses. Or, there may be a buildup of ammonia fumes in the building, which can cause respiratory lesions, infection, and subsequent chronic drainage. A rare rabbit may even have sniffles because of an allergy to certain things in the rabbitry such as straw or the powdered limestone that is sometimes used to keep floors dry. Constant exposure results in continuing irritation of the respiratory system.

Symptoms: A wet nose is usually the first indication that a rabbit has caught a cold. The rabbit may sneeze or cough; it may also wipe at its nose with the front feet frequently. A rabbit sometimes sneezes for no apparent reason yet does not develop a cold. However, if a certain rabbit is seen sneezing frequently, watch it for further symptoms.

Effect: A sudden light cold usually seen in a freshly kindled doe during inclement weather can slow down her milk production if it is allowed to go untreated. The doe may suffer a lack of appetite, which in turn affects her ability to provide milk. The nursing doe is the hardest hit by colds, due to the fact that her system is already under a certain amount of stress.

Contagion: A rabbit is not, under good conditions, prone to colds. However, once a rabbit has caught a cold, it should be treated quickly to prevent a worse problem and the likely spread of the cold from that rabbit to the entire herd. The rabbit should be isolated from contact with animals in adjoining cages. For rabbits are by nature sociable critters and often touch noses through their wire cage walls. Thus, the cold of one rabbit is passed on to the next cage, and from that cage on to the next one, and so on.

Treatment: Once a rabbit has caught a cold, the best-known treatment is penicillin, given by injection according to the rabbit's weight and size. For a mature medium-sized breed of doe (New Zealands, English Spots, and Tans for example) that weighs from 10½ to 12 or 13 pounds, an injection of 2 cubic centimeters of penicillin once a day for three consecutive days usually cures a cold. When giving penicillin to smaller rabbits, weigh the animal to be treated, and measure the dosage accordingly. For instance, a dwarf breed such as the Black Dutch or a young rabbit that weighs less than 6 pounds should receive only ½ of a cubic centimeter of penicillin per injection. A rabbit weighing from 6 to 9 pounds can *normally* handle an injection of 1 cubic centimeter of this antibiotic. (I stress *normally*, since there is an outside chance that a rabbit may suffer a sensitivity to penicillin. Since the penicillin-sensitive rabbit is a rarity and there is no known method of predetermining sensitivity, you simply have to take the small risk. If the rabbit is left untreated, the cold is likely to advance and may eventually kill the rabbit.) Giving penicillin to rabbits younger than six weeks is not usually required and is not recommended.

Penicillin should be used with caution. First, the real need for antibiotics should be determined; too large a dose can kill, and too small a dose may do more harm than good. Second, if penicillin is given for a sudden cold or other infection, it should not be continued for more than three consecutive days. Prolonged use of antibiotics can result in an immunity to the beneficial effects of these medications. Also, adverse changes in the vaginal secretions of does may occur which cause sterility, and lack of interest in breeding.

Prevention: Preventing colds in the rabbitry is not as difficult as it may sound. One of the simplest and most economical methods of prevention is the culling of stock that seem to have any weakness for catching cold. That is, a doe that repeatedly catches colds every time she kindles a new litter obviously is more inclined to have colds than the doe that has litter after litter with no colds.

A good sound feeding program can go a long way in preventing run-down conditions which can lead to colds. Vitamins in water during bad weather help. A great many breeders in recent years report that they have eliminated problems during bad weather by regular feeding of comfrey. Comfrey feeding acts more as a preventative when used on a daily routine basis than as a cure after a rabbit has a cold.

Prompt and thorough protection from drafts is insurance against colds. Draft-free, adequate air circulation is essential, since this helps to prevent a buildup of ammonia gases emanating from urine and droppings beneath cages. Making certain that any fans used do not blow directly on the rabbits is another way of insuring rabbits against colds.

Since damp, chilly living quarters cause a lot of colds, using fans in the rabbitry when the building must be closed keeps the air circulating, which helps decrease moisture. Proper drainage in the soil where the rabbitry is located is a tremendous help in keeping down moisture inside the building. Regarding a chilly rabbitry, rabbits can stand more cold than they can heat, but they cannot tolerate drafts or living in damp cages or having to sleep on damp straw. The use of heat in the rabbitry, except in the most extreme cases, does more for the comfort of the breeder than for the rabbits!

As for air circulation during mild months, do not worry about the balmy breezes blowing throughout the rabbitry. Domestic rabbits thrive on open air as long as it isn't terribly cold or wet. It is the draft coming in through a crack or slit in the wall that does the damage in causing colds. Even in the coldest weather, rabbits love *indirect* flows of fresh air. The one exception is the naked newborn baby which has only the nest straw and hair for protection against the temperature.

Diarrhea

There are several types of diarrhea which can afflict rabbits. Young rabbits between the ages of five and eight weeks are the most susceptible.

The most common kind of diarrhea stems from one or more of the coccidiosis germs. This type may be no more than a tendency toward softness in the droppings, or it may be a profuse, usually dark-colored bowel discharge. At times the fecal matter may have a somewhat watery appearance. Young rabbits are most commonly troubled with coccidiosis-based diarrhea, but rabbits of any age may be affected at times. Those afflicted usually display a lack of appetite, resulting in an insufficient gain or a loss of weight.

Occasionally, you will see a young rabbit that has considerable soiling of the rear, but the rabbit apparently feels frisky, eats well, and seems to keep up with its littermates in weight gains. The soiling is usually a dark greenish black color and has a mealy consistency. This is suspected to be merely a case of the little rabbit making a pig of itself at the feeder. It overeats, then when the dry food is being digested, there is a greater demand for more water. So the rabbit drinks an excess of water. The stomach forces the food through the intestines only partially digested, which results in the mealy appearance in the fecal material. When the rabbit is given a little straw to eat, this condition ordinarily disappears within hours.

Bacterial diarrhea is another matter altogether. This type can be recognized by the rank, sour smell emanating from the vicinity of the rabbit. Droppings are usually dark, soft, and sticky. The rabbit suffering this type of scours should be isolated, and medication is recommended.

Treatments for bacterial scours range from chemical medications to old-fashioned elm or plantain leaves. Sulfa-based drugs such as those used for coccidiosis are usually effective. For people wishing to stay with more natural medications, the ill rabbit may be given a handful of either fresh green elm leaves or green plantain leaves. You may also give it a piece of freshly cut elm branch, leaving the bark intact on the limb. A piece of limb only about four or five inches long is usually given to a rabbit or a doe with a litter that is suffering bacterial diarrhea. Results with either the chemical or the home remedy are likely to take a day or two; in other words, the rabbit may show some recovery within one day, but if the medication is discontinued sooner than three or four days, the ailment is most likely to return.

Fresh blackberry leaves may also be used for the bacterial type of scours. Under no circumstances should ill rabbits be given more green stuff than they will clean up within one hour, since the stale, wilted greens aggravate the intestinal problems, especially in young rabbits.

Mucoid enteritis is one of the most dreaded types of bacteria diarrhea, since science has never been able with any certainty to isolate the organism responsible. Nor has anyone been able to come up with an absolute or a natural cure. Many remedies appear to work in some cases but fail completely in others. This is true of both chemical and natural methods of treatment.

As the name seems to imply, mucoid enteritis is a passage of mucous-like stools. There may be some brownish staining in the droppings, but the major portion of bowel movements will have a mucoid consistency—a clear or whitish jellylike material. The rabbit with this disease usually has a pinched look about the face. Its spine can be felt, thin and ridged, and the belly will have a pot appearance since it will be badly bloated, as a rule.

Left unattended, the rabbit usually dies within a few days, but may linger on for a week or more in some cases. Rabbits suffering from mucoid enteritis rarely eat anything, but usually appear to have an enormous thirst. Therefore, treatment is most effective when given in drinking water. But due to the general dehydration of the body, oral medication alone is normally insufficient to prevent the rabbit from succumbing.

Be prepared for a disappointment in trying to save the mucoid victim. In fact, there may be more disappointments than successes. However, if you are inclined to try to save each and every victim, some results may be obtained by giving the rabbit an intramuscular injection of 5 cubic centimeters of a sterile saline solution. The inside of the heavy part of the hip is recommended for giving shots of any kind, but the injection may also be given on the outside of the hip. Avoid hitting the bone with the tip of the needle, which may be difficult on the small rabbit that is already dehydrated to some extent. The saline injection only prevents further dehydration; actual treatment via the drinking water consists most commonly of Terramycin®. Give this drug for at least five days, or until the rabbit returns to normal eating habits. In some few cases, livestock calcium administered in drinking water at a ration of one teaspoon to one quart of clean water has been known to save early victims. But this, like all other known treatments for mucoid, works only some of the time—at other times it does no good whatever.

Mucoid enteritis seems to be a tenacious problem, and discontinuing treatment too soon usually means a recurrence of the illness. Mucoid is not believed to be contagious, but isolation of the victim is recommended, to be on the safe side. Many longtime raisers report that while mucoid is not considered contagious, cases of it appear to increase the more they try to work with and save rabbits already afflicted. Therefore, experienced breeders generally dispose of a single victim rather than risk further herd contamination.

Most experienced raisers feel that, without undue stress on the herd from weather changes or something else, inherited weaknesses seem to cause a great portion of the problems with diarrhea. This is not to say that the diseases themselves are inherited; a weakness of a certain type in family bloodlines can allow alien organisms access to the intestinal tracts of rabbits stemming from those family lines. Therefore, it is wise to choose replacement breeding stock from only those litters that have no problems at all during their early, most rapid growth periods. A litter which has had quite a lot of intestinal trouble while growing out is quite likely to produce in turn young animals which also have numerous problems.

Sanitation in the rabbitry is one of the best preventatives of diarrhea. Since many troublesome types of diarrhea hit herds hardest during very damp, chilly weather, it is helpful to make sure that the rabbitry is located on well-drained soil, and that there are not water or urine puddles in the rabbitry. Fresh uncontaminated water is a must. Food should be kept in sealed storage facilities where dogs, cats, rats, and mice cannot contaminate it; metal or plastic garbage cans with lids are excellent for this storage.

If there is a problem with water or urine puddles under cages, fresh clean sawdust makes a good absorption agent while it also adds a fresh aroma. Wood chips are also used in some rabbitries to help keep down odors and to soak up wet spots.

Prompt medication, whether chemical or natural, is the surest way of preventing the spread of diarrhea. A rabbit should never be allowed to wear out the ailment; more often, the ailment will wind up wearing out the rabbit and killing it! Treatments for simple scours that are simple and easily applied by anyone are: plenty of good clean straw, preferably oat if it is available, and fresh clean water twice a day. For bacterial scours: sulfa-based medication in drinking water, strictly according to directions on the package; isolation from the herd; sanitation in the rabbitry and within the hutch; and a little clean dry straw to eat; or, all the plantain or elm leaves eaten within an hour; a piece of elm branch with bark left on; and sanitation just as with the chemical medication.

Pregnancy Problems

While a doe is not normally prone to pregnancy or kindling problems, an occasional doe may have trouble.

One of the most common problems does have in kindling is heat prostration. Pregnant does are much more susceptible to hot weather than humans. When does are close to their delivery dates, during very warm summer days, the wise breeder stays pretty close by. If at all possible, he keeps the does in the coolest, airiest parts of the rabbitry.

If you see a pregnant doe lying stretched out flat on her belly, head raised high, and panting for breath with her mouth open, move quickly to cool her. This can be done by placing a wet (thoroughly soaked) burlap or paper feed bag under her, or by giving her a pad of thoroughly soaked straw to lie on. But given wet straw or a bag, she may carry it right into the nest box when she is ready to kindle; the wet material will chill the naked newborn babies, killing them in most if not all cases.

The *safest* method of cooling a too-warm pregnant doe, is to place her in a floorless cage directly on the cool ground. Here she will usually stretch out full-length on her belly. She may move from time to time, seeking a cooler place to lie. The nest box can be put right into the cage with the doe; this gives her something familiar and reassuring. It also ensures that the doe will deliver her young in the nest instead of scattering them over the cool earth, should her time arrive while she is resting there.

Probably the next most common pregnancy problem is ketosis. This is also sometimes called the fat doe disease or pregnancy disease. It is a disturbance in a doe's metabolism, usually caused by a too-generous feeding program. The pregnant doe should *never* be given all she will eat. She will ordinarily have a prodigious appetite if she is in good health. A medium-sized doe like a New Zealand should never receive more than six ounces of solid food; a Dwarf or Dutch doe that weighs five pounds or less should get three ounces. The pregnant doe may be given a few leaves such as plantain, a carrot once or twice a week, or a teaspoon once a day of a protein supplement such as Calf Manna®. Feedings may be divided so that the animals receive half of their daily ration in the morning, the other half in the late afternoon. But never should the daily allowance of solid feed exceed six ounces per day.

The doe may act hungry when you go near her cage, but you should not give in and provide another feeding. Offer her a handful of good clean oat straw if you haven't the heart to pass up a hungry pregnant doe. The straw may be given in addition to the few fresh green leaves, if desired, since there is little food value in the straw. One of the major benefits of feeding straw, aside from filling the rabbit, is improved digestion from the roughage. Few does that receive small daily rations of straw during pregnancy will suffer either constipation or loose bowels.

An occasional doe may have kindling trouble because of too-large babies. This most often occurs when the raiser has bred a very small doe to a large heavy buck. The offspring may then inherit the sire's size, making it difficult, if not impossible, for the small doe to deliver them. This is why experienced breeders always suggest breeding animals that are fairly well matched in size, and using only animals in the same breed range; that is, never use a New Zealand buck to breed a small Dutch or Dwarf doe.

Short of calling in a veterinarian, there is little the breeder can do to help the doe trying to kindle too-large babies. Since vets are rarely available on less than several hours notice, this is usually not practical. It seldom happens when the breeder uses breeding bucks of the same breed and general size range, so it is rarely a problem for the small rabbitry.

Losses of baby rabbits at or immediately after delivery stymie some breeders. Often there is a simple explanation. If the doe has kindled a part of her litter on the cage floor and they survived, you may find babies crawling around on the floor halfway across the rabbitry. Of course, if hutches are outside and there are cats or dogs on the homestead, then chances are the babies will never be found. Rats, mice, and snakes are known to steal babies from nests if these are not kept under control. On rare occasions, the doe herself may eat one or more babies when they are first born. When the doe does this, though, there will usually be bits and pieces of the babies lying either in the nest, on the cage floor, or below the cage on the rabbitry floor.

A doe's body sometimes, though rarely, absorbs the partially formed embryos. Ordinarily, this only occurs when the doe is subjected to extremely stressful conditions, either from nutritional deficiency or badly crowded quarters (such as a cage or hutch that is totally inadequate in size). Unless the doe is receiving a truly small ration of food daily, this absorption does not mean that her ration should be increased in weight. More often, the problem lies in a lack of nutrition in the feed provided.

A failure to conceive may or may not be the fault of the doe in question. Look over the breeding record of the buck with which she was mated. Check how many litters he has sired against how many does he has bred within a two-week period prior to and following the mating with the doe that failed to conceive. If there are other does that have missed after being bred by the same buck, the buck may be worked too often and simply be a bit run down. During very hot weather some bucks tend to go sterile (some does also, but not as often as bucks), and thus may cause a number of does to miss, though they may be quite willing to go through the motions of breeding. One good practice during summer is to check the buck before allowing him to mate. If he is fully fertile, there's a good chance that his testicles will be full and smooth looking. Should the testicles have a shriveled, shrunken appearance, it is better to mate the doe with another buck or wait a few days.

If one doe in particular seems to misconceive on a regular basis when other does bred by the same bucks kindle nice litters, then the doe is at fault and should be culled from the herd. A sterile doe will eat just as much as a fertile, high producing one! In fact, it might even be said that the infertile doe costs more than the fine production doe: she is not only taking up cage space, eating as much as a pregnant doe every day, taking up the raiser's time in feeding and care, but she also is taking away from the good doe's production by not repaying even her own keep.

Even where normally adequate nutrition is provided, occasionally rabbits will suffer a shortage of vitamin E when weather conditions are bad and the herd has been producing well over a long period of time. Adding a tablespoon of wheat germ, or wheat germ crumbles, or a few drops of wheat germ oil to the daily feed ration can often correct the situation. It does not do anything for the habitual misser, but works wonderfully on those rabbits that have formerly been good workers.

Breeding Problems

Spring is the normal breeding time for all rabbits—wild or domestic, cage-raised animals.

During extremely hot or cold weather, the raiser may have some difficulty in getting usually eager breeders to mate. This is sometimes due to sterility in either or both mates. In some cases, I suspect that their reluctance may be due to an innate sense that mating during weather extremes would produce a litter at one of the most dangerous times of the year: when temperatures are so high that the doe would suffer heat stroke while kindling, or when the weather is so cold that the young may freeze when they are first born. (Many people working with rabbits after a while become convinced that animals are smarter than people when it comes to their native, inborn instincts!)

Natural wheat germ added to the daily ration often helps spur the rabbits to breed in any season of the year. Some breeders advocate putting a doe in a cage next door to a buck, allowing them to see each other and touch noses. Other raisers claim good results by using massage: rub with the fingertips around the flanks and hips of the buck; and around the neck, under the chin, and along the belly of the doe. There is one problem with this practice: if the buck or doe is sterile at the time, it is encouraged to go through breeding, with another misconception to show for the raiser's trouble.

Just occasionally, you may come across a doe that is a perpetual nonbreeder. That is, she consistently refuses to have anything to do with a buck, or at best seems interested in breeding only now and then. In such a case, the raiser is wise to cull the doe and replace her with one from a family line of willing breeders.

Since sterility is the basic cause of breeding problems and usually a result of weather extremes or imbalances in the diet, the raiser has to use his own judgement to determine which of the two is creating his breeding trouble.

Where a nutritional deficiency is the culprit, this can be easily corrected by providing a more balanced diet. You have less control over the weather, but you can often create an environment more conducive to regular and willing breeding. Keep breeding bucks and pregnant does in the coolest, airiest section of the rabbitry; it can help tremendously when summer rolls around.

If outdoor hutches are used, provide a good dense shade over the hutches without obstructing the fresh air flow; this helps keep rabbits fertile and in the mood for regular mating.

Do any breeding chores very early in the morning during hot summer months; better yet, do them during the evening hours when the sun has been down for some time. The reverse is best when winter comes along: handle breeding chores during midday when temperatures are highest.

A doe may have special problems when she has been nursing heavily to feed a large litter. Always check a doe for general condition before breeding her again after she has nursed a large litter for six or seven weeks. If the doe appears terribly thin and has a rough coat and her eyes lack their usual brightness, then a portion of the litter should be weaned. Only two or three bunnies should be left with her to nurse (to prevent caking of the breasts); the doe and the remaining bunnies should still receive full feeding.

As a rule, within a week to ten days of this special feeding, the doe's condition will show marked improvement. If not, you may have to resort to extra special feeding and provide the doe additional flesh-building nutrients. Whole rolled oats can help the nursed-down doe regain her usual good condition. It doesn't take much; a tablespoon or so of oats per day will do the trick.

As soon as the doe in question has gained back some of the lost weight, she may be returned to a buck for mating. Leave the bunnies with her for a few more days, until it becomes obvious that her milk supply is diminishing, then wean them and let her rest until she is due to kindle, thirty-one days after breeding. When the young are all weaned, the doe's ration must be cut back to the five- or six-ounce-per-day feeding of solid food. She may be given small amounts of green stuff and straw.

Experts claim that rabbit does do not have heat cycles in the usual sense. However, many breeders have found there are times when their does simply refuse service by bucks. When this happens, some recommend what is called force breeding. This means you hold the doe in the mating position. Grasp the doe's ears and the loose hide over the back of the neck and shoulders, slide the other hand back under her belly to between the back legs, then elevate her hindquarters so that the buck may mount. Many longtime raisers have compared this method with natural breeding and discovered that where natural mating gave an 80 percent or better conception rate, the forced breedings gave only about a 20 percent success rate.

The breeding method used most often is to check the doe for readiness. Pick up the doe by the loose skin over the back of her neck and shoulders. Bracing her back against your thighs, with the fingers of your other hand, press in on either side of the vaginal vent. If the vent is a pale pinkish color, there is little chance that the doe will willingly accept the buck. If the vent has a dark reddish or purple coloring and appears slightly swollen, the doe is quite likely to accept service quickly.

A few raisers make a habit of leaving a reluctant doe in the cage with a buck for several hours, even several days when trying to keep up their production ratio. In some instances this might work, but in others it results in problems larger than the benefits.

Unwilling, angry does have been known to partially castrate fine, aggressive working bucks when left in the same cages with them for long periods of time. Does that are not this aggressive may accept service by the buck on more than one occasion as much as several days apart. When this happens, the doe just may have multiple litters born several days apart. The first litter is naturally kindled in the nest box and nursed as well as possible while the doe's body is busy building a second litter.

The second litter may also be kindled in the nest box but is more likely to be delivered on the cage floor. A second litter, smaller and weaker than the older litter, has no chance to obtain the milk they need, since the first group will push them away from nipples. The second, weaker litter will starve—sometimes quickly, sometimes slowly. I have known does to kindle two litters as much as two weeks apart, then abandon the second litter because she already had a litter in the nest. These litters came from test-breedings done two weeks after the does were first bred. The does delivering these second litters were among those rare does that accept a second service even though they are already pregnant from a previous mating.

Sore Hocks

Sore hocks (the large back feet) are not too common a problem. Since the usual cause is inherited thin pads on the hocks, it is a difficult ailment to cure in most cases. However, conscientious breeders down through the years have done their best to eliminate this hereditary factor.

The large back feet of any future breeder should be thoroughly checked. If the fur pads are not extremely thick, the rabbit should never be kept, as it is quite likely to have sore back feet. As a rule, mature stock with very thick, heavy pads on the hind feet will pass this trait along to their offspring.

A well-padded rabbit rarely has trouble with sore feet. When one does, check first the cage floor. A broken wire there can cause wear and tear of the fur on the soles. When the fur is worn through, the bare skin is subjected to constant irritation. If a broken wire is found, it can be bent downward so that the protruding tip does not come in contact with the rabbit's feet.

Sometimes the sore feet are caused not by a broken wire but by a rough rusted floor. Rusted spots can often be repaired without replacing the entire floor. Using a stiff wire brush, clean off as much of the rust as possible. Then, making certain that a nontoxic aluminum paint is used, repaint the rusted place. Paint can be either sprayed or brushed on. Allow it to dry completely before returning the rabbit to the cage. To prevent additional rust spots, repaint the entire cage floor when you make these repairs; this adds further life to the wire.

Another cause of sore hocks is an accumulation of wet hair, urine, and crushed feces in one or more corners of the cage. This keeps the rabbit's feet damp, resulting in constant irritation to the skin on the bottom of the feet. The fur pads begin to come out and leave unprotected patches that become raw. Frequent cleaning prevents the buildup of debris. The hair that rabbits shed has a tendency to cling to any wet surface, even the narrow surface afforded by wire in the floor. This hair may be burned off of the wire on a regular basis, or it may be brushed or washed off. Brushing is recommended only as a stopgap method of removing hair from cages; in itself it does not kill any germs. Brushing does prevent accumulation of soiled hair in cage corners, thus assisting germ control to some extent. But, for the real cleaning of a hutch, burning or washing with a disinfectant solution is suggested. If hair burning is practiced with a small torch, the wire should be completely cool before the rabbits are again housed in the cage. When the cage is washed, all moisture should be allowed to dry before returning the animals. Either of these cleaning methods helps to keep disease germs under control, also.

While special medications can be bought for sore feet, the best remedy I ever used was ordinary carbolated petroleum jelly mixed with flowers of sulfur—both available at most drugstores. Combined in a ratio of 2 parts jelly to ½ part flowers of sulfur, these ingredients heal the injured part of the foot. Used properly, the resulting ointment also helps pad the worn-through spot until the fur has regrown. Neither ingredient will harm the rabbit, should it lick some of the ointment. (No one has yet come up with a method of preventing rabbits from licking medication off any part of their bodies!)

A number of breeders report excellent results in healing sore hocks by using udder ointments (sometimes called bag balms). Obtained at veterinary supply stores or your local feed mill, the udder ointment is made for milk cows' or goats' udders when they have suffered wire or nail cuts or punctures.

Hutch (Urine) Burn

The common cause of hutch or urine burn is almost exactly the same as wet-corner-created sore hocks—a buildup of hair, urine, and soil on the cage floor. When a doe sits on the wet dirty area, chemicals in the urine-soaked mat burn the tender skin of the external vaginal vent and surrounding area.

Sometimes a doe causes urine burn by consistently wetting in her nest box. When the nesting material is soaked with urine, the doe may sit in the wet straw for hours at a time and thus acquire a dilly of a burn.

To prevent does from suffering these burns, the same cleaning procedures should be used as just discussed under Sore Hocks. Nest boxes should be checked daily to see that none are being used as bathrooms by the does.

If a doe persists in urinating in her nest box, it should be removed immediately when the wetness is discovered. The doe should be provided with a fresh clean box with clean, dry nesting material. Spraying or daubing the box with Lysol® seems to help discourage does from urinating there, but it does not interfere with their instinct for kindling in the box. Merely replacing the wet nesting material while leaving the urine-scented box appears to encourage the doe to go right on using the box for a bathroom. (Using the damp box can also endanger—because of germ growth—the newborn bunnies kindled in it.

To cure hutch burn where there appears to be no broken skin or infection, simply wash the area with a mild soap, then rub on an antiseptic ointment or lanolin, or the petroleum jelly and flowers of sulfur ointment mentioned under Sore Hocks. Treatment should be repeated every day until the skin has returned to its normal healthy appearance. Eliminating the cause is essential to prevent continuation of the problem.

Where there is broken skin and/or infection in the urine-burned area, first wash the area with a mild soap or antiseptic to remove external bacteria. Working very gently, press out any accumulation of pus and remove it with sterile cotton; you may not be able to remove all of it. After pressing out any pus, rewash the area with a mild antiseptic, then dry gently with cotton or absorbent paper that is known to be clean. Treat with an ointment for ordinary hutch burn; repeat this daily until the infection and burned look have completely disappeared.

During the period of treatment for urine burn, whether a mild or severe case, the animal should never be bred. Hutch burn itself is not contagious, but there may be secondary bacterial factors that can be picked up by another rabbit during breeding.

Eye Problems

Most problems with eye infections are found in very young rabbits. The troubles stem from several sources: the mother doe urinates in the nest; there is excess dust in the nesting material; a little rabbit, while crawling and squirming around in the nest, has its eye stuck by the end of a straw; or the litter is exposed to cold or damp conditions that give one or more bunnies a slight cold.

Normally, eyes are open between the twelfth and fourteenth day after birth. Each bunny in a litter should be checked, beginning on the twelfth day, to make sure its eyes are opening properly. If one or more bunnies have an eye that is not opening while others have eyes partially or completely open, keep a close watch on those unopened eyes.

If, by the fourteenth day, the eyes of the entire litter are not opened, very gently open them with the fingertips. Once each eye is open, an antiseptic eye ointment may be used to get rid of any germs trying to gain a foothold. There are special ointments and powders for use in these cases. However, where there is no apparent infection (usually visible purulent matter in or around the eye), a good eyewash is usually enough to prevent a reclosed and possibly infected eye.

Whether treated with an ointment, powder, or eyewash, a bothersome eye should be checked daily to make sure it doesn't reclose. As a rule, once an eye is opened and treated for a day or so, it will remain open, clear, and bright. If an unopened eye remains untreated, it can often result in a total loss of sight in that eye.

To prevent eye infection in the tiny bunnies, make absolutely certain that nest boxes are thoroughly cleaned and sterilized before they are used. (See page 39 for nest box sanitation.) Another prevention is to check nest boxes every day to make sure that does aren't soiling the nesting material. If a nest is found wet or badly soiled, a freshly sprayed box should be prepared and given to the doe. To prevent her disowning her young in a different box, salvage whatever clean, dry nesting material possible and transfer it to the new box, especially any of the doe's fur that is still dry and clean. This gives the fresh nest at least a little of the doe's own scent. She isn't likely to resent the fresh Lysol® scent but may object to the new straw and box that do not have her own scent on them at all.

Eye infections in adult rabbits usually are caused by ammonia build-up in the rabbitry; colds stemming from exposure to cold drafts; or injury from some sharp object in the cage—a broken wire end, the corner of a feeder, or a chewed-off place on the edge of a nest box. On rare occasions, a doe may fight with a buck at breeding time and stick a claw into his eye, creating an infection.

The treatment for adult eye problems is the same as that for the very young animal: clean the eye, then treat it with ointment or eyewash. If the problem stems from a cold, penicillin injections for two or three days may be necessary. Where the infection comes from an injury, penicillin may also be necessary to promote the healing process. In any event, clean and treat the eye daily until all signs of infection are gone; otherwise, the infection may recur.

When excess dust causes eye problems, several things can help. First, nesting material: try to store it where there is the least possibility of dust accumulation. If you use straw, you can sometimes find a farmer whose storage methods decrease the dust in his straw. As a last resort, shake out handfuls of the straw to dispose of as much dust as possible before putting the straw into the box.

If dust from outside the building is a problem during dry summer months, a screen of burlap bags hung on the windward side can help enormously. It is amazing how much dust a burlap awning can accumulate in a few weeks—dust that would otherwise drift right on through the rabbitry! Wetting down the ground outside during dry weather helps diminish dust problems, too. If the interior of the rabbitry is very dry, spreading clean sawdust on the floor helps to keep dust from flying while you work around the building.

Since a great many eye troubles, in both young and adult stock, stem from a buildup of ammonia gas in the rabbitry, prevention of this through good sanitation is most important. The basic method is to simply clean out the manure regularly, at least once every eight to ten days. Spreading sawdust or fine wood chips over the entire floor immediately after cleaning keeps down odors and prevents any buildup of ammonia in the air; besides, the sawdust lends its own aroma to the air for several days. Further prevention of eye infections depends on elimination of cold drafts and broken wires.

Abscesses

The occasional abscess on a rabbit can have more than one cause. The abscess can start from a small puncture wound in the skin. It can come from a fly that lays its eggs just under the skin. Or, it may stem from *Pasteurella*—the most dangerous and difficult source to control.

Any sharp item protruding into the cage can puncture the skin of the rabbit when it is romping around. Bacteria may gain access to the wound, causing an infection and pus. To prevent such an injury, simply eliminate an offending object from the cage.

Treatment is usually as follows: first make a small slit on the lowest side of the abscess with a very sharp instrument such as a razor blade, press out and remove the pus, then wash out with an antibiotic. It's useless to put a bandage over the wound, since a rabbit will most likely lick and pull the bandage off. So, to protect against reinfection, an antibiotic ointment may be applied thickly to the area. The rabbit will probably lick off most of the ointment, but it assures some measure of protection. Check and treat the wound daily until healed. Most breeders advise shaving all fur around the puncture before attempting treatment.

The swelling that is caused by the larvae of a fly (of the variety *Cuterebra cuniculi*) is not actually an abscess in the true sense of the word. Rather, the swelling is created by the growth of eggs deposited under the rabbit's skin by the adult fly. The eggs grow into larvae which in time will exit from the flesh as flies.

To treat the affected rabbit, first determine if there is larvae growing under the skin. Part the fur over the swollen area. Look for a small opening inside which you can see the tip-end of larvae. Disinfect the surrounding area, then shave off all hair in the immediate vicinity. Using sterile cotton, gently wipe away the larvae and as much pus as possible. Then treat the area with an antibiotic powder or ointment. Clean and retreat it each day until completely healed. It is not ordinarily necessary to reshave the area, since the hair will grow back slowly enough to allow total healing. Without daily observation and follow-up treatment, though, the wound could become reinfected.

Pasteurella-based abscesses are more difficult to heal once they have started. These are caused by a general environment of bad husbandry in the rabbitry and a resultant lowering of the rabbit's natural resistance to disease-causing organisms. Under stressful conditions that decrease the native stamina, *Pasteurella* organisms gain access to the rabbit's system. Abscesses are only one of the external manifestations of an increase of these germs. (See Nutritional Deficiencies, page 115.)

When an abscess is believed to be caused by *Pasteurella*, it may be opened and treated in the same manner as the abscess caused by a puncture wound. The animal should be given intramuscular injections of an antibiotic such as penicillin. Since the multiplication of *Pasteurella* organisms is most commonly due to some situation that is stressful to the rabbit, measures must be taken to correct this.

Caution: Take extreme care in handling an abscess of any sort, since some bacteria which cause it are contagious and can spread to humans and other animals. Wear rubber gloves if at all possible when treating an abscess, burn all cotton swabs used in removing pus, and be sure to clean and sterilize all instruments used in opening an abscess. If rubber gloves are not immediately available, be sure to scrub your hands and change into clean clothes after working with an abscess, before handling or working around other animals.

Isolate the afflicted rabbit from other stock until all signs of abscess are cleared up. Clean and disinfect any cage or equipment used by the abscess victim.

Chewed Fur

Most hair chewing is done by young rabbits, but adults also are sometimes guilty of this irritating habit. First signs of fur eating usually show up around the faces and heads of a litter, and sometimes on a strip down the back.

Some raisers blame hair eating on a lack of sufficient fiber in the diet. Others claim that sheer boredom—being cooped up in a hutch with little to do but eat and drink—causes the rabbits to start chewing fur from littermates or themselves. However, considerable research indicates that the majority of hair eating stems from a dietary deficiency. The research departments of major feed companies have found that increasing the protein content of rabbit feed eliminated or greatly decreased the incidence of fur chewing. Our own experience also verified that giving more protein, either in the form of a protein supplement pellet or a type of feed with a higher protein content, decreased or eliminated any fur-chewing problems we encountered.

People who feed additional alfalfa hay or comfrey rarely have problems with fur chewing. This would, of course, take the place of a protein supplement pellet in most cases.

Occasionally, a doe will eat all of the hair she pulled to bed her new babies, or she may even chew all the fur off her young when they have furred out and come out of the nest. Again, additional protein can usually stop the doe from practicing this bad habit.

Hair is almost completely protein, so it appears logical that other sources of protein would be a cure for hair chewing. In some instances, additional roughage in the diet might help.

Like humans, some rabbits need more of a certain nutrient than others do. So, those rabbits with a consistent appetite for hair may have nothing more than a protein shortage in their normal diet. (See Protein, in chapter 4.)

Caked and "Blue" Breasts

The two most common causes of caked breasts in rabbits are: an injury from a sharp object that causes a bruise in the mammary gland, and the loss of a baby in the litter of a heavily milking doe.

When exercising in her cage, the doe may hurt her breast on a corner of the feeder or any sharp, rough place on her nest box. Usually a slight swelling of the nipple shows at first, followed by reddening of the injured area and hardening of the nipple. You'd be wise to check freshly kindled does every day or two; rub your bare hand along the full length of the belly to spot any unusual heat or solidity of a nipple.

In the early stages of injury-caused caked breast, the doe may be treated by simply rubbing camphorated oil or Campho-Phenique® into the damaged nipple. Treatment should be repeated daily until all signs of swelling are gone. The oil won't harm the doe if she licks it off. Neither does it appear to harm the babies, though they may have oily looking little faces from nuzzling into the oily fur. The camphor scent does not discourage the young from nursing heartily.

If a bruised nipple is allowed to go untreated, it may turn black or dark blue. In severe cases, the fevered breast may start to crack and eventually come off completely, leaving only scar tissue. When used at an early stage, the camphorated oil prevents this drying out until the swelling has disappeared. The camphorated oil tends to dry up the milk in the treated breast but won't affect milking ability from other nipples. Care should be taken not to spread oil on unaffected nipples; it is possible to dry up a doe's complete milk supply by reckless spreading of camphorated oil!

When an extremely heavily milking doe loses one or two babies out of a large litter, quite possibly those spare nipples will not be needed to fill the remaining babies. Within a day or two after losing a baby, accumulation of milk may begin in an unneeded nipple. Soon the breast becomes feverish and the unused milk starts to harden (cake). Left unattended, the doe will be in real trouble within a couple of days. Treatment with camphorated oil also works well in this case.

Besides an injured nipple or a lost baby, caked or blue breast may stem from an increase of those tricky little *Pasteurella* germs. Some does kept in dirty, damp, drafty quarters regularly suffer from this problem. In these cases, the diagnosis is probably *Pasteurella*; apply medication and sanitation practices accordingly.

Treatment with the camphorated oil will help to some extent in these more severe cases, but it won't cure the basic cause of *Pasteurella*. This calls for antibiotics in whatever dosages are prescribed by your veterinarian.

To prevent ordinary caked breasts, be sure there are no sharp objects in the doe's cage. Check nest boxes for any chewed-down sharp places, nails, or tacks that may be protruding. Fix any bent corners on metal feeders. When the heavy milker loses a baby, check her breasts every day for at least three days for an unusual heat, swelling, or hardening where milk may not be nursed out.

To prevent the more severe blue breast, does with injuries or lost babies should be checked and treated faithfully until symptoms disappear. While *Pasteurella* seems to be a dormant factor in many rabbits, the problems caused by this germ can easily be kept under control by proper sanitation, ventilation, and a well-balanced diet. Most does suffering blue breasts from *Pasteurella* infections have likely been exposed to stressful conditions over a period of time. Their resistance has been continually beaten down by a situation that is detrimental to their health, and it may take some time to correct.

Cannibalism

On rare occasions, a doe may eat part or all of her newborn litter. Once the bunnies are a few days old, cannibalism is even more rare. When there are no bits and pieces of babies left in or under the cage, suspect first some outside source for any loss; some raisers blame a doe for babies lost that instead have been stolen by rats, snakes, or other animals.

As with hair eating, additional protein in the diet usually prevents a doe that eats part of her litter from repeating the act on another litter. This is not to claim that all cannibalism is caused by protein shortage or can be stopped with protein. In some very rare instances, a doe seems to insist on destroying part or all of every litter rather than caring for it. Not all animals make good mothers, just as not all human females make good mothers!

Outside disturbances sometimes cause a nervous newly kindled doe to eat or kill her litter. A dog or cat prowling nearby can really create havoc in a freshly delivered doe. In a rare instance, the presence of strange people in the rabbitry causes a doe to kill her young. When gentle calm-natured does are saved for breeding stock, this does not usually happen. Extremely nervous, high-strung, easily frightened does are the worst for killing their young. Sometimes an uncommonly severe storm with loud thunderclaps can frighten a normally calm doe and cause her to leap in and out of her box, stamping the young and killing them.

A doe is instinctively quite fierce in protecting her newborns from dangerous intruders. Where rats or mice are allowed to inhabit the rabbitry, you can expect a doe to sometimes trample her young while fighting to protect them. A snake in or near her cage can send the calmest doe into sheer hysteria, even though the snake may be only a harmless garden variety.

Now and then you will find a partially chewed-up baby in the cage or nest box of a first-litter doe, due to her lack of experience. She is naturally nervous and isn't really sure just what she is supposed to do with the new baby, but instinct tells her that the cord must be cut and the baby cleaned. In biting the umbilical cord in two, the nervous young mother bites too deep, cutting into the belly of the baby rabbit. As a rule, only one such dead baby will be found, as the doe instinctively corrects her mistake on subsequent babies delivered. The young doe should not be blamed for this accident. If she later eats others of her first litter, then provide her an additional ration of protein. Should she eat part or all of her second litter, then cull her from the herd.

Ear Mites

Ear mites can lead to problems such as infections in the ears, lack of condition, loss of appetite, and lack of interest in breeding. The mites that cause these problems are soil-living creatures, microscopic in size. Being invisible to the naked eye, mites can become a real threat to rabbit health without the raiser even suspecting there is any trouble, unless he keeps a frequent watch on his animals' overall appearance.

The first indication of ear mites is seen in tiny red spots or specks on the tender inside of the ear. Left unmolested, the mites proliferate so quickly that within a matter of days, the rabbits' ears may be severely infested. Advanced cases will cause brownish-colored scabs to form. Infection can form under these scabs. The instinctive scratching of the

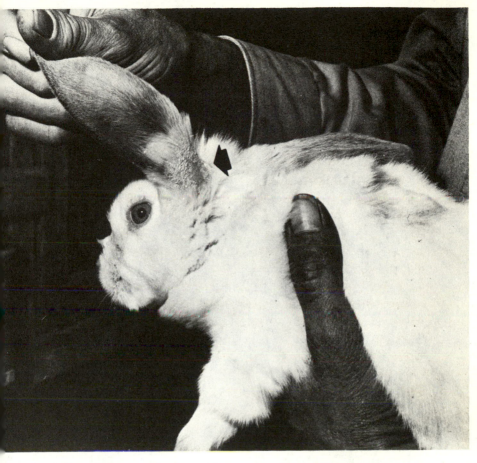

Ear mites aren't the only cause of ear ailments in rabbits. The rabbit above is afflicted with fur loss and scaling skin caused by ringworm. (See also page 152.)

ears by the animal only further irritates the sensitive skin and spreads the infection.

Medications for the prevention of ear mites can be purchased in veterinary supply stores or through animal or pet journals. However, these special (and more expensive) medications are not really necessary for the small breeder.

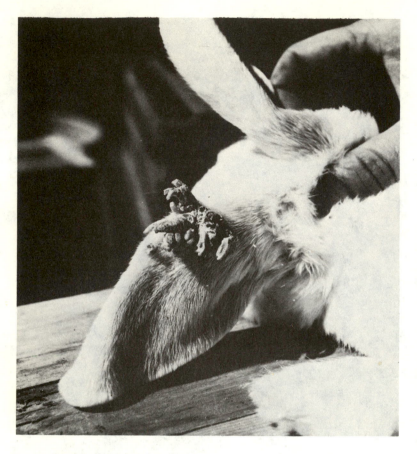

Another ear ailment is this growth caused by a papilloma virus.

An excellent (and inexpensive) medication can be made by combining five parts of heavy hospital-grade mineral oil with one part ordinary camphorated oil or better yet, because it mixes more easily, Campho-Phenique®. Mix the medication in one of those small plastic squeeze bottles with the pointed cut-off tip. This serves as a perfect applicator, making it very easy to drop exactly the required amount of medicine into each ear. In cases of advanced infestation, repeated treatments may be necessary. For routine prevention, just two or three drops in each ear once a month is usually adequate.

Some raisers advocate the use of a few drops of iodine mixed with mineral or olive oil. Having experienced iodine on a few scraped elbows and knees as a child, I wouldn't advise the use of it inside a rabbit's ears; the stuff burns like fury, even on a toughened knee!

Other breeders use just the oil, without any additive. Cooking oil may be substituted if mineral or olive oil is not readily available. Any light oil will work, but it should be as colorless as possible to prevent staining of the animals' pelts. The oil will control the mites through suffocation if there is no infection, but it will do little to heal any small lesions caused by the colonies of mites burrowed beneath the skin.

Paralysis

Paralysis is normally considered to come from some spinal injury suffered while the rabbit is exercising in its cage. The rabbit rarely recovers from such an injury. Paralysis may also result from the organism *T. gondii*, discussed in the section on Stress-Related Bacterial Diseases. This type may be associated with or preceded by emaciation, lack of coordination, and tremors in the nervous system. Ingestion of the wooly-pod milkweed, most commonly found in the southwestern area of the United States, may also bring on paralysis.

In any case of paralysis, regardless of cause, the most humane treatment is to put the poor animal out of its misery. Otherwise, it will suffer a prolonged and painful degeneration due to its inability to reach feed and water. Also, the stress of the paralysis allows secondary disease organisms easier access to any part of the afflicted animal than would otherwise be possible. No type of medication is recommended, since it will at best prolong the rabbit's suffering.

"Wobblies"

One rare problem—sometimes diagnosed improperly as rickets—occurs only in very young rabbits. In this ailment, the legs are normally straight and strong looking. Compared

It isn't always simple to accurately diagnose an ailment. This rabbit's wry neck is caused, not by some skeletal deformity, but by a middle ear infection that has upset its sense of equilibrium.

to rickets, the back legs do not splay out from the body, which would cause the rabbit to walk on its front legs and drag the hindquarters along. Rather, when the rabbit moves around, the head may wobble from side to side (thus the name "wobblies"), the front legs may suddenly stumble as if the rabbit had tripped, or the whole body may simply flip over to one side or the other. In trying to get up when it falls, the rabbit may wallow around and kick its legs. Usually it can eventually fumble its way to feed and water but may fall over while trying to eat or drink. In a few isolated instances, the rabbit may suffer a slight diarrhea and soil its hindquarters.

Until four to six weeks of age, the rabbit appears normal compared to littermates. The young animal may continue to grow at a rate very near that of others in the litter, but it usually shows signs of weight loss within a few days after the staggering is first evident. In some cases, the rabbit even reaches fryer size, though it may be a little thinner than other fryers.

In more severe cases, the little rabbit becomes totally disabled—unable to control its movements sufficiently to eat or drink—and finally starves. When its skull is cut open during an autopsy, the brain sac is found to be a grayish, dull, opaque color rather than the pinkish color of that in a healthy young rabbit.

This disease is thought to be caused by too-close inbreeding, which results in various degrees of a sort of mental retardation in the young. In fact, some raisers call rabbits showing these signs of partial incoordination "little idiot rabbits," since their actions often apparently lack intelligence. Other breeders, when they are unlucky enough to produce little rabbits with these symptoms, call them "wobblies."

It is easier to dispose of afflicted rabbits early when symptoms are first noticed than it is to wait and butcher them as meat, unless you are totally unaffected by these mentally deficient animals and their apparent complete lack of fear and obvious affection for whoever furnishes their feed and water. They respond differently to humans than normal rabbits do.

Chapter 6
Using Rabbits

Slaughtering and Dressing

There are probably as many variations for killing meat animals as there are people raising and butchering them. No *real* animal raiser does a slow, inhumane, sloppy job of it, though. I doubt that there are very many who actually enjoy killing and dressing out an animal. I have never met people who relished this task, though I have known those who went about it in a matter-of-fact manner, knowing they were doing the cleanest, most humane job of it possible.

One of the most popular methods of killing a domestic rabbit for meat is to strike one swift, sharp blow at the base of the skull. Depending on your force and accuracy, the rabbit will be killed, knocked unconscious, or merely stunned. Always try for an accurate blow, hard enough to kill with just one lick.

Working rapidly, insert a sharp knife through the skin between the tendon and the flesh just above the back knee joint of the rabbit. This makes a slit in the back leg that is quickly slipped over the end of a gambrel (meat hook, also sometimes called a hanging hook).

With the rabbit securely hung on the gambrel, slit the neck just behind the chin line or jawbone. A cut farther back decreases the size of the usable pelt. Working with the sharp knife-edge, cut through the cartilage of the spinal column, thus severing the entire head. Many people save the head; they cut through the skull and collect the brain for cooking in various ways.

Next cut only the skin around the back legs just below the gambrel and make a single cut across the skin of the

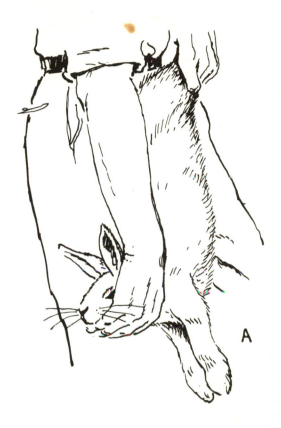

A

Getting the rabbit ready for the freezer is never pleasant. One way is to knock the rabbit out by dislocating its neck, as shown in A. Then hang the rabbit by its hind legs from a gambrel, cut its throat, then bleed it out. Remove the head and forepaws (B). Then cut through the hide at the hocks and begin working it off the body (C). When the hide is free of the hocks, pull it down over the body (D).

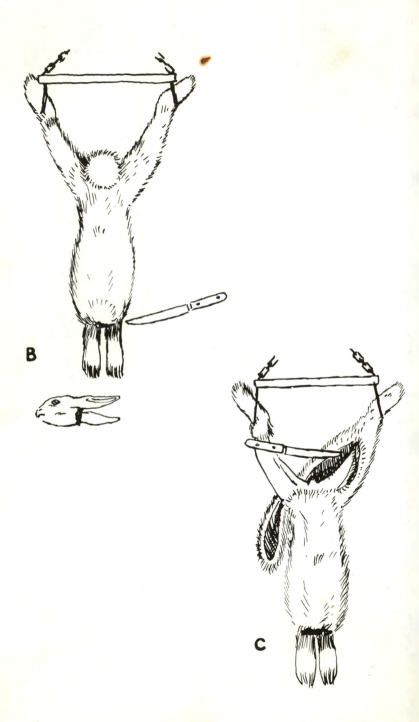

B

C

belly from one leg to the other, to the cuts just below the gambrel hooks. The flesh is thin in this belly area, so take care to prevent puncturing an intestine.

Still cutting through only the skin, make another cut up from the belly slit, around the genitals, anus, and over the upper side of the tail. (The tail may later be removed and cured for use as a decoration on children's garments, if desired.) Now, remove the two front feet at the first joint.

With these basic steps finished, fully loosen the skin from the back legs. (Speed in this operation comes from practice.) Then using both hands, pull the hide evenly downward toward the head. This is called casing, or taking the pelt in one piece with no side cut open. The skin will be inverted as it comes off the carcass. During this skinning operation, the carcass will bleed out to a great extent.

The next step is the removing of the intestines. With a very sharp knife that has a keen tip, continue the cut that was made earlier around and above the tail, down to the rear end of the rib cage; be extra careful to cut through only the thin flesh of the belly. Slide two fingers down into the rib cavity, then place the blade of the knife between the fingers and the rib bones. Work the knife through the cartilage which connects the rib bones to the breastbone, thus separating them and opening up the entire carcass for removal of the intestines.

Being careful not to overstretch the intestines, pull them loose from the interior of the carcass. In a few places it may be necessary to use the knife tip to cut through a bit of supporting tissue. Before the intestines are completely removed, you may want to cut the liver loose from them. On one side of the liver lies the gallbladder, which must be removed. Leaving the gallbladder intact on the liver, or accidentally breaking it and spilling gall upon the liver or other parts, will spoil the meat and give it a very unpleasant, bitter taste. The heart may also be cut loose from the insides at this time and saved. Some people claim that the liver, heart, and brain of domestically raised rabbit are the real delicacies, and they save these parts until enough are collected to make a meal of them.

Once the heart and liver are out, finish removing the intestines from the carcass. What is done with the now-skinned and gutted carcass depends on the preference of the raiser *and whether or not he intends to sell any rabbit meat to others*.

Many people who butcher rabbits *only* for their own use drop the carcass into ice water to finish bleeding and cooling it. Others place the carcass in a pan or tub of ice, which accomplishes the same end.

If you intend to sell rabbit meat to others, find out beforehand just what is required under your state laws that pertain to retail sales of meats butchered at home. Some states are fairly permissive, as long as the meat is butchered in the cleanest, most sanitary manner possible. There are states, though, that can enforce pretty strict meat inspection laws. Some states permit the sale of home-butchered meats, providing the meat is cooled in a manner that does not use ice water. Others require cooling by refrigeration and forbid retail sales of home-butchered meat cooled in any other way.

When preparing fresh meat for one's own table, the cooling method is up to the person doing the butchering. No one can tell the raiser that he *must* butcher his own meat for his private use in any certain way—at least, not yet!

When the carcass has cooled sufficiently for easy handling (usually not more than fifteen or twenty minutes if ice water or ice is used), start the cutting-up procedure by removing the legs. The front or the back legs may be used as a starting point, but most people I've watched have started with the forelegs.

Lightly pull the whole leg away from the carcass, and with a sharp knife slice down from the underside of the leg until contact is made with the shoulder joint. Give the joint a slight twist to show exactly where the cartilage is. Cut through the cartilage, then on through the rest of the shoulder meat. Repeat the same procedure on the other shoulder and then on the thighs. Unlike a chicken, the lower leg and thigh are not usually separated on a rabbit carcass. Both front and hind legs are served as single pieces.

Cut through the flesh which separates the rib cage from the lower back. Then, grasping the entire back in your hands, bend the spine backward until the cartilage shows between the vertebrae. Cut through the cartilage. This makes two large pieces from the back of the carcass. With a sharp knife (a sharp-pointed boning knife works well here, but a thin butcher knife blade will do the job), slice along the gristle or cartilage that joins the ribs to the backbone on either side of the spine, making two pieces of the rib section.

The rear section of the back may be served as one large piece, or it may be separated into two pieces like the rib portion was. This section—sometimes called the loin or the saddle—is one of the meatier pieces of the rabbit. When the cutting up is finished, you will have either seven or eight pieces of delicious white meat (depending on whether the loin is served as one piece or two), plus the liver, heart, and possibly the brain.

Approximate Yield: The yield of meat per fryer depends on the type of rabbit raised, how the rabbit is fed, and whether or not the fryer has been a victim of some minor ailment before reaching the usual four- to five-pound slaughtering weight. It also depends on the bone structure of the breed of rabbit raised. A large-boned rabbit, such as the giant breeds, will provide less meat per pound compared to the amount of waste (bone) than a medium-sized breed.

Ordinarily, a young rabbit between the ages of eight and ten weeks will have a yield of between 2 and 2½ pounds of fine white meat. The usual commercial type of fryer will give approximately a 50 percent dressout ratio. That is, 50 percent good meat, 50 percent bone and other waste. Some few raisers have claimed as high as a 60 percent dressout for their fryers, but the average fryer, well-fed and healthy, will yield between 50 percent and 55 percent meat.

The shorter bodied, "cobbier" type of fryer tends to yield more meat than the longer bodied, lean and racy type. The meat of either type is equally edible and equally nutritious.

Older rabbits, from twelve weeks of age on up, are also quite edible. These rabbits normally weigh from six to ten or twelve pounds each. The younger of these can be used in almost any recipe calling for either younger or older rabbit meat. The "chunkier" the body type, the more meat one gets per carcass, naturally. Older rabbits need to be boiled or baked, rather than fried, as the younger ones are used.

Even the doe or buck that has ceased to function efficiently as a breeder can be either baked or boiled. The meat can be used in soups, casseroles, or stews. Many people prefer the older rabbit over the young fryer, claiming that the meat has a much fuller flavor when baked with vegetable stuffing.

Nutritional Value: More progressive hospitals serve domestically raised rabbit meat to convalescents, certain ulcer victims, and to those having digestive problems that limit the types of meats their systems can handle.

According to the United States Department of Agriculture, rabbit meat offers the highest per pound content of protein of seven meats served in America.

Rabbit meat offers 20.8 percent protein per pound, compared to 20.0 percent for chicken, 18.8 percent for medium fat veal, 20.1 percent in medium fat turkey. A good grade of beef contains 16.3 percent, medium fat lamb offers 15.7 percent, and pork comes in last with only 11.9 percent.

Domestically produced rabbit meat contains less fat. Comparative figures are: rabbit, 10.2 percent; chicken, 11.0 percent; medium fat veal, 14.0; medium fat turkey, 20.2; good beef, 28.0; medium fat lamb, 27.7; pork loses the battle against fat in the diet by a wide margin, 45.0.

Moisture content in meat offers no nutrition at all. Here again, rabbit meat leads other meats with the smallest amount of moisture per pound. Comparison with other meats shows: rabbit meat, 27.9 percent; chicken, 67.6 percent; veal, 66.0 percent; turkey, 58.3 percent; lean beef, 55.0 percent; lamb, 55.8 percent. Pork is rabbit meat's nearest competitor in moisture content with 42.0 percent.

When it comes to calories per pound, rabbit meat once again comes out ahead. Rabbit meat contains 795 calories per pound, with chicken running the closest second with

810 calories. Veal is next in line with 910 calories per pound. Turkey, surprisingly, contains 1,190 calories per pound.

Lamb is next lowest on the calorie content chart with 1,420 calories per pound. Coming in next to last is that good old standby in our American diet, beef, with 1,440. And, those pork chops and hams that so many of us consume with such relish contain a whopping 2,050 calories per pound.

Rabbit Meat Recipes

As mentioned previously, rabbit meat can be prepared in so many ways that collecting all the recipes into one book is a virtual impossibility. In addition, most chicken recipes can be used in cooking rabbit, though remember that rabbit meat is very mildly flavored, and strong spices or vegetables in chicken recipes may overpower the flavor of the meat.

Frying is the most common method of cooking the young rabbit from eight to ten weeks old. In some cases, rabbits up to twelve weeks of age can be cooked this way. The portions are treated much the same as those of a frying chicken. Dredge each piece in lightly seasoned flour, then cook in shallow hot oil or fat until golden brown on all sides.

Fryer rabbits may also be steam-fried for a juicier, richer tasting dish. When all portions have been browned as described, drain off as much oil as possible, then cover the skillet with a very tight-fitting lid. Lower the heat and steam the rabbit for fifteen to twenty minutes. (If steam escapes from the pan, add about one tablespoon of water to prevent sticking and scorching.) Remove the lid and cook for another five to ten minutes for a crispy crust on the meat.

Remove the meat and sprinkle into the pan about two tablespoons of flour, cornstarch, or arrowroot starch flour. A teaspoon of butter or margarine may be added if extra shortening is required to saturate the flour. Stir in water (or milk, if preferred) to make a gravy from the flour-shortening mix. Two tablespoons of flour require about 1½ to 2 cups of liquid. Serve with a light salad, a vegetable, and a drink, for a delicious meal.

The steam-frying method may be modified to make what in the South used to be called smothered meat. Instead of removing the browned rabbit from the skillet before making the gravy, simply pour off excess fat or oil. Then add water till the meat is just barely covered and maintain at a slow simmer. Simmer the meat in the resulting sauce until the gravy reaches the desired consistency. If a thicker sauce is required, mix 1 teaspoon of flour or starch with ½ cup of water; add this slowly to the simmering meat and thicken as desired.

Many rabbit raisers in the South are quite fond of a dish that probably stems from the Chicken 'n' Dumplings that is so well known in the area.

Rabbit 'n' Dumplings

1	large frying rabbit, cut into serving portions
¼	cup finely diced hearts of celery
¼	cup finely diced onion
1	tablespoon butter or margarine
½	teaspoon salt
⅛	teaspoon white pepper
	water to cover

Combine all the ingredients in a pot which holds approximately ⅓ more than the volume of the ingredients.

Cover with a tight-fitting lid and cook slowly until the meat is fork-tender. While it is cooking, mix the following to make the dough:

1	cup white flour
2	teaspoons baking powder
½	teaspoon salt
1	teaspoon butter or margarine
	water or milk to moisten
	to a drop-biscuit consistency

When rabbit meat in the large pot is tender, drop teaspoonsful of the dough into the simmering liquid. Keep the dumplings separated in the water as much as possible, to prevent them from sticking together. As soon as all of the dumplings have been added, cover tightly and raise the heat until the liquid comes to a rolling boil. Boil covered for ten minutes. Remove cover, lower the heat and cook at a low simmer for ten minutes. (To make more dumplings— for a large family or company dinner—use two cups of flour instead of one, and add one more teaspoon of baking powder plus an additional ¼ teaspoon of salt.) Makes six servings.

Rabbit and Biscuit Pie

1	cut up rabbit
	(either a fryer or roaster)
½	cup chopped celery
¼	cup finely diced onion
1	tablespoon butter or margarine
½	teaspoon salt
¼	teaspoon white pepper
1	cup cooked green limas
	or English peas
1	cup cooked diced carrots
1	cup water mixed with
	1 tablespoon flour or cornstarch

Combine above ingredients in a casserole large enough to allow two inches of free space above the mixture. Cover tightly and bake at 350°F. for one hour, or until the meat is tender.

Biscuits for crust:

2	cups flour
½	teaspoon salt
3	teaspoons baking powder
½	teaspoon sugar
½	teaspoon baking soda
2	teaspoons melted shortening
	sour milk or buttermilk

Combine the dry ingredients. Add the shortening and enough sour milk or buttermilk to make a stiff dough. On a floured board or cloth, roll out the dough to about ½ inch thickness. Cut small biscuits, about 2 inches in diameter.

Remove casserole from oven and, while the liquid is still bubbling, place the small biscuits over the top. (For added richness, tops of the biscuits may be brushed with melted butter or margarine.) Return to the oven uncovered, and continue to bake until biscuits are browned, usually from fifteen to twenty minutes.

Makes six servings.

This same recipe may also be used for Rabbit Pot Pie by removing the bones from the precooked meat and substituting rolled crust in place of the biscuit crust.

Rabbit with Rice

6	slices breakfast bacon
	salt and pepper
1	cup flour
1	frying rabbit,
	cut into serving portions
1	medium onion finely diced
2	bouillon cubes
2	cups hot water

Fry the bacon until it is crisp. Remove from pan and allow to cool. Salt, pepper, and flour the pieces of rabbit, then fry in the bacon drippings until browned. Remove the pieces and place in a casserole with a lid. Sprinkle the diced onion over the rabbit; crumble the bacon and sprinkle this also over the rabbit.

Dissolve the bouillon cubes in the hot water and pour over the rabbit, cover, and bake at 350°F. until the meat is tender. Remove to serving platter.

Thicken the gravy with flour or cornstarch and serve over steamed rice. Garnish with parsley.

Makes six servings.

Vegetable-Stuffed Rabbit

1	whole roasting rabbit
	(twelve weeks of age or older)
1	tablespoon salt
½	teaspoon white pepper
1	cup coarsely chopped onion
1	cup sliced carrots
1	cup cooked green limas
	or English peas
1	cup coarsely cut-up celery
3	quartered large potatoes
½	cup finely diced green pepper
½	cup water
¼	cup melted butter or margarine

Rub inside of rabbit cavity with combined salt and pepper. Mix all vegetables together, adding leftover salt and pepper. Stuff the cavity lightly with some of the vegetable combination. Arrange remaining vegetables around the meat in a roasting pan with a cover. Add the water and brush the upper side of the meat with the melted butter or margarine. Bake at 350°F. until meat is fork-tender, brushing occasionally with the butter or margarine to prevent drying of the meat.

Remove stuffed rabbit to a serving platter. Transfer extra vegetables to a separate serving dish. Add hot homemade bread, salad, dessert, and a drink for a fine large family or company meal.

Serves from six to eight.

Rabbit Sandwich Spread

1	cup precooked, cooled, boned, and ground rabbit meat
½	teaspoon salt (or salt to taste)
	mayonnaise or salad dressing to moisten

Mix thoroughly. (If a more highly flavored spread is desired, add a dash of celery salt, white pepper, or seasoned salt.) Spread on very lightly buttered bread or toast. Toasted bread seems to enhance the delicate flavor of rabbit meat.

Charcoal-Barbecued Rabbit

	2- to 3-pound young rabbit, cut into serving portions
1½	teaspoon salt
¼	teaspoon white pepper
½	cup sherry
½	cup cooking oil
1½	teaspoon seasoned salt

Sprinkle moist pieces of rabbit with salt and pepper and place on rack over a medium-hot bed of coals. Make a sauce by blending together the sherry, cooking oil, and seasoned salt. Turning the pieces often, baste frequently with the sauce. Cook for one hour, or until tender. (A favorite barbecue sauce may be substituted, but additional oil may be necessary to help keep the rabbit tender and juicy.) Makes four to six servings.

Rabbit Salad

2	cups cooked, boned, and diced rabbit
½	cup finely diced celery
½	teaspoon salt
¼	teaspoon paprika
	mayonnaise or salad dressing to moisten
	large lettuce leaves
	parsley

Combine rabbit meat, celery, salt, paprika, and mayonnaise. Serve in a mound on crisp lettuce leaves. Garnish with parsley. Serve with toast points or crackers.

Creole Rabbit

½   cup flour
1½   teaspoon salt
¼   teaspoon pepper
     2- to 3-pound young rabbit,
        cut into serving portions
¼   cup butter or oil
     creole sauce

Combine flour, salt, and pepper in paper bag. Add moist pieces of rabbit to bag, and shake until all are well-coated with the flour mix. Heat the butter or oil in a heavy skillet and brown meat lightly on all sides. Transfer the browned rabbit to a casserole dish. Pour the creole sauce over rabbit, cover and bake at 325°F. one hour or until tender. Uncover and bake thirty minutes to brown the top. Makes four to six servings.

Creole Sauce:
2   medium sliced onions
1   clove minced garlic
1   tablespoon chopped parsley
1   tablespoon butter or oil
3½   cups tomato juice
¼   teaspoon Worcestershire sauce
     salt and pepper

Cook onions, garlic, and parsley in fat until onion is golden brown. Add tomato juice and Worcestershire sauce. Simmer for 15 minutes and season to taste.

Green Peppers with Rabbit Salad

4	medium green peppers
2	cups diced cooked rabbit
½	cup fresh peas
½	cup sliced carrots
½	cup chopped celery
1	tablespoon grated fresh onion
1	tablespoon chopped fresh parsley
1	tablespoon snipped fresh chives
¼	cup mayonnaise
¼	cup sour cream
¼	teaspoon curry powder
½	teaspoon salt
2	teaspoons fresh lemon juice

Cut off tops of the green peppers and remove the seeds. Cook peppers uncovered in boiling water for five minutes, drain and set aside. In a large bowl combine the rabbit, vegetables, and herbs. Mix in a small bowl the mayonnaise, sour cream, and remaining ingredients. Add this to the rabbit mixture and mix thoroughly. Fill green pepper shells with the rabbit salad, and chill until ready to serve. Makes four servings.

Rabbit and Tortillas en Casserole

1	frying rabbit
1	package soft tortillas, broken into pieces
1	can tomatoes with hot peppers

1	chopped large onion
1	chopped large green pepper
	red pepper
	salt
1	can cream of celery soup
1	can cream of mushroom soup
1	cup broth
	grated cheese

Cook rabbit and remove bones. Cut the meat into small pieces. Grease a flameproof 7½-by-11½-inch pan; line it with the broken tortillas. Combine half of the tomatoes with the onion and green pepper. Arrange the meat over the tortillas, and sprinkle with red pepper and salt to taste. Pour the tomato mixture over the meat, and add the soups and broth. Simmer for thirty to forty minutes. Top with grated cheese. Bake at 350°F. until the cheese melts.
Makes six to eight servings.

Oven-Fried Rabbit

1	frying rabbit,
	cut into serving portions
¼	cup melted butter or margarine
	salt and pepper

Brush pieces lightly with melted butter or margarine. Sprinkle with salt and pepper to taste, and place in a flat baking dish. Cover with foil (or a tight-fitting lid). Bake at 350°F. for one hour or until tender, occasionally brushing with melted butter or margarine until lightly browned on top.

Barbecued Rabbit—Oven Style

	2- to 3-pound frying rabbit, cut into serving portions
1½	teaspoon salt
¼	teaspoon pepper
¾	cup barbecue sauce (your own or bottled)
½	cup water

Sprinkle moist pieces of rabbit with salt and pepper. Place in a shallow baking dish and brush generously with barbecue sauce. Pour the water in the bottom of the dish, cover, and bake at 350°F. for forty-five minutes. Remove the cover, turn the pieces, and brush them well with barbecue sauce. Bake uncovered for thirty minutes, or until well browned; brush occasionally with more sauce to keep moist.

Makes four to six servings.

Rabbit Stew

1	cup flour
1	teaspoon salt
½	teaspoon white pepper
1	stewing rabbit (6 pounds or over), cut into serving portions
¼	cup butter or oil
½	diced onion
½	cup chopped celery
	water
1	cup green limas
1	cup sliced carrots
3 or 4	quartered large potatoes
1	cup English peas

Combine salt, pepper, and flour in paper bag. Shake rabbit portions in bag until coated with the flour mix. In a heavy deep pan, heat butter or oil. Add floured meat and cook until brown on all sides. Drain off excess fat. Add onion and celery and enough water to just cover. Cover with a very tight-fitting lid to prevent escape of steam and flavors. Simmer until meat is tender. Remove from heat, take out rabbit pieces, and remove bones. Return the meat to the pot and add the limas, carrots, and potatoes. Cook over low heat until vegetables are fork-tender, then add peas and continue to cook until peas are just tender. Season to taste with additional salt and pepper if desired. (This is one of those dishes that should not be tackled in a hurry; it takes approximately two hours from start to finish and must not be hurried. Cooking at too high a heat can cause larger pieces of rabbit to be tough and stringy.)

Rabbit Livers and Hearts

1	cup flour
½	teaspoon salt
½	teaspoon baking powder
	water
1	pound rabbit livers and hearts
	deep fat for frying

Combine flour, salt, baking powder, and enough water to make a thin batter (about the consistency of thin pancake batter). Dip individual livers and hearts into the batter. Drop into 350°F. deep fat; fry until the batter turns a golden brown. Remove and drain on paper toweling or a clean tea towel. (To make a rich batter, add one beaten egg.)

Pelts

Pelts are left alone until the carcass has been dressed out and refrigerated.

To obtain furs that last a long time—especially for larger, frequently used things like coats—it is best to either send pelts to a professional tanner or obtain one of the several books on home-tanning of pelts and carefully follow its directions.

When pelts are intended only for trim or small fad items for a child, the skins can be cured at home without complicated tanning processes. The following are simple home cures for hides that will not be subjected to hard or constant wear. These hides will not be as serviceable as professionally tanned pelts.

One of the simplest methods I've heard about, is to rub fresh skins generously with a half-and-half mixture of salt and alum on the flesh side only. Dry in a cool shady place for three days, after which the skins should be stone-dry.

Carefully pull off the dried fatty tissue, then trim the edges of the pelts to make straight sides. Wash the skins by hand in a mild soap or detergent and rinse out the suds. *Never twist or wring a pelt!* (The skins may be washed in a washing machine, but hand washing is preferred.)

Hang the wet hides—with the fur side on the inside— over the clothesline in a shady spot. Sunlight dries skins too rapidly, which causes stiffness. As the pelts dry, pull them back and forth across the line several times to loosen and fluff the fur. They may take a couple of days to dry completely. After drying, rub neat's-foot oil or saddle soap into the skin side to soften the pelts. Several rubbings may be required to make the skins really soft.

A slightly more complicated method of curing raw pelts is to rub each skin generously with salt. Then put the skins into a tub or other container. Cover the container and leave for four days outside or in a building where there is *no heat.* Remove the skins from the tub, and scrape the skin side with a dull knife or piece of broken glass until all of the fat and gristle are off. Discard any residue left in the container.

Next, make a solution of 2 pounds of salt, 1 pound alum, and 2 gallons of water. If you are treating only a small number of furs, reduce the amount of curing solution to 1 pound salt, ½ pound of alum, and 1 gallon of water. Put the skins into this solution for two weeks.

After two weeks, remove the skins and wash them in mild soap or detergent, then hang them over a line in a cool shady spot. As the pelts start to dry, pull them back and forth across the line from time to time to loosen and soften the fur. This method requires about two days of drying time. The fur should be checked for dryness down next to the hide before starting the rubbing to soften the skin. If the fur is not completely dry, it may tend to mat when handled.

Again, rub saddle soap or neat's-foot oil into the skin side to soften it. Once the skin is softened, these pelts may be used for small items such as caps or hoods for children.

Rabbit furs can be used in numerous items: bedspreads, patchwork coverlets, lap robes, cushion covers, pajama bags, handbags, neckties, stuffed toys, hats, caps, hoods, mittens, baby booties, vests, coats, capes, clothing trim, and so on. In fact, if you can make an item out of other materials, chances are you can also make it of rabbit furs.

Preparation and Uses of Rabbit Wastes

Rabbit manure is one of the finest fertilizers a gardener can obtain. Unlike other animal manure, you can use rabbit droppings directly from the rabbitry, and they will not burn even delicate young plants.

A ten- to twelve-pound doe with only twenty-eight young per year together pile up approximately six cubic feet of clear (uncomposted) manure per year. Depending on the moisture content, one cubic foot can weigh up to forty pounds. Thoroughly air-dried, this weighs about sixteen pounds; fresh manure weighs approximately twenty-eight pounds per cubic foot. Before it is mixed with other ingredients in a compost pile, rabbit manure contains about 89 percent organic matter, 6.5 percent water, 2.3 percent nitrogen, 1.3 percent phosphoric acid, and 0.7 percent potash.

Rabbit droppings can be composted naturally by stocking the compost pile with earthworms, or more quickly by using an activator. We always used the natural method, which gave us ready-to-use compost in about three weeks during warm months.

In the moderate temperatures of the Ozark Mountains, we used an aboveground compost pile. By simply setting posts to form a square area about four-by-four feet, we made a pit six feet deep and enclosed it with ordinary poultry mesh. This pit provided sufficient space for composting the manure from an average of 110 does with their offspring, plus from fifteen to twenty-one mature working bucks.

To start the first batch of compost in the pit, I put down a layer (about six inches deep) of used nesting straw taken from nest boxes being cleaned. On top of this I added a shallow layer of damp soil, then several inches of rabbit droppings. Seed beds of large earthworms were placed into the droppings. More nesting straw was placed over the worm nests, and then the pile was sprinkled with water from a garden hose.

Be careful not to make the pile too wet; otherwise the earthworms will drown. Keep the composting material just damp. Make a cover over the compost pile, if desired, by spreading a sheet of black plastic over the top and weighting it down with rocks, pieces of scrap lumber or wood, or strips of heavy metal. (Just use whatever is most convenient—even a few tree branches, if these are handy.) However, if the climate is not experiencing heavy rainfall, it is not really necessary to cover an aboveground pit. We rarely used a cover and never had problems with worms drowning in rainstorms. A dug pit sunk into the earth presents more problems with water accumulation and should have a cover at least during rainy periods.

Another method of building compost pits near a rabbitry is to use concrete blocks to make low walls with drainage holes near ground level. Screen these holes to prevent worms from migrating from the walled-in pit to other areas. Some worms migrate downward and away in any case, but nowhere near the entire colony does.

In our own rabbitry buildings, we always maintained worm beds under the hanging wire cages, besides in the compost pile. When manure was cleaned from under the cages, no precautions were taken to prevent removal of some worms from there to the compost pit. It takes only about ten days for mature worms to replenish the supply of egg capsules under the cages.

Many products on the market are supposed to keep livestock housing odor-free. Most of these products are chemicals—some of which will destroy all of your earthworms when they are used. But there is one deodorizing product that is natural, harmless, extremely effective, and usually either free for the asking or quite inexpensive. Furthermore, it can be recycled into compost when it is no longer effective within the rabbitry.

An elderly, highly experienced rabbit raiser introduced us to the use of ordinary sawdust as a rabbitry deodorizer. Sawdust has a natural absorbency, so that it soaks up most of the moisture that accumulates under rabbit cages, which otherwise would be ideal fly-breeding areas.

The only tools necessary are a shovel (if you have the muscle power to handle a heavy scoop of sawdust), or a one-gallon hand scoop. The hand scoop can be either purchased or made in the same manner as a feed scoop out of one of those gallon-size plastic bleach bottles.

To obtain the best results from the combined use of earthworms and sawdust, spread a thin layer of sawdust immediately after cleaning the manure from under the cages. With worms working the droppings, cleaning usually won't be necessary more often than once every two weeks.

Without worms, clean once a week—once every eight days at the very outside; during very hot humid weather, clean at least once a week if not more often, to keep down odors and flies.

Some raisers claim that if earthworms in large numbers work the droppings under cages, it is not necessary to clean there more often than once every six months. However, a shallow pit (about ten to twelve inches deep) under a cage can hold only so much material, be it whole droppings or worm castings! An individual raiser may be able to clean less often than the two weeks which I have suggested. We found that with our herd, cleaning every couple of weeks was most satisfactory; this kept any droppings from rolling out into the aisles where I had to walk. Maybe I was a bit picky, but I never did like walking around in manure—not even in those round little rabbit droppings!

When both worms and sawdust are used under the cages, both help minimize odors. The worms work the droppings into what are called worm castings, which are odorless. Should a wet spot develop under certain cages, it can be turned and stirred with a manure or potato fork, and fresh sawdust can be spread on that particular area to soak up excess moisture.

Besides deodorizing by soaking up excess urine in one spot under a cage, fresh sawdust adds its own fragrance to the rabbitry atmosphere. We always preferred to get hold of some good fresh pine sawdust, since the pine has such a delightfully refreshing aroma. But any sawdust that is fresh and dry lends its own particular scent. All sawdusts absorb moisture under the cages equally well.

There has been some conflict about using sawdust on gardens or flowers, because it supposedly depletes nitrogen supplies in the soil. However, sawdust that has been used as an absorber under rabbit cages does not have this effect on garden soil! The rabbit urine counteracts it.

When worms and sawdust are both used on the manure, and the combination of sawdust and manure is transferred to the compost pile where more worms work on it a while, there will be no sawdust evident in the compost. Everything becomes a fine, mealy residue ready for use on even the tenderest young plantings. The residue (worm castings) may be worked into the soil, increasing the water-holding qualities of the land, or it may be mulched around plants. Worm castings and leaf mold are often used as potting material for house plants.

In every rabbitry the soiled nesting material must be disposed of in one way or another. Some raisers advocate burning the straw or hay (or wood shavings, shredded cane stalks, or whatever is used). However, homesteaders can use an infinitely better way of disposing these wastes. Instead of piling the dirty straw outdoors and sticking a match to it, dump the bedding (including the doe's hair) on the compost pile. Use it as a layer between layers of manure. When all soiled nesting material has been spread over the top of the pile, dampen it down with a water hose. Adding a layer of manure on top of the straw hastens its disappearance, since earthworms in the compost will attack the fresh eatables with vigor as long as the straw is slightly damp and offers them a dark protected atmosphere.

Dust found in pelleted feeds, or even in crushed or ground natural grains, can also be used to advantage in the compost pile. Most of the metal feeders we used had small holes in the bottoms through which the feed dust sifted and fell to the worm pits below the cages. The dust added to the food available for the worms. Left in feeders, the dust can cause respiratory problems when rabbits inhale it.

Even the hair in soiled nesting material is disposable in the compost pile. Apparently earthworms devour the stuff; normally hair is very slow to decay.

When you have a compost pile going strong, with thousands upon thousands of those friendly little worm critters working for you, even the occasional lost mature rabbit can be disposed of by burying it deep down in the pile. The fur will disappear, as well as the bones in time. Earthworms have obviously voracious appetites! And the buried baby rabbit or two lost from a litter will disappear within a few days. But I would not advise the disposal of a great number of large rabbits this way. I never had occasion to compost more than one adult at a time, and I suspect that disposal of a number of large rabbits would present some odor problems. Normally, the worm-worked compost pile poses no trouble with odors whatever.

In addition to the disposal of rabbitry wastes and the occasional rabbit carcass, the worm-worked compost pile is the ideal homestead garbage disposal unit. We deposited *almost* everything in our compost pile: table scraps, canning wastes, spoiled or wet hay and straw, weeds, grass cuttings, tree trimmings, waste paper, garden wastes such as spoiled vegetables, vegetable peels, fruit rinds, and so on. I say *almost* everything, since there is one thing that must never, under any circumstances, be put into a compost pile where earthworms live: grease or oil! Any kind of cooking grease, shortening, or oil will kill the worms.

As for the number of worms you need to start under-cage dropping pits and a compost pile: When we first introduced worms into our rabbitry, we already had nearly fifty working does, and we obtained two standard galvanized washtubs full of worms and bedding material. In rows that were about fifteen feet long, we spread the two tubfuls in the pits below the cages. Actually, you don't spread the worms. It is better to scoop out a hole every so often (ours were about three feet apart under the fifteen-foot rows), put two large handfuls of worms and bedding into the hole, and cover them over with a rake or shovel.

You won't notice much for the first few weeks. Then suddenly (it seems) the droppings will start to look different. You won't observe so many whole droppings— mainly only those from the past day. Upon raking or shoveling away some of the surface, you will see a fine, mealy substance. This is proof that your worms are doing their duty.

At first the difference in the material will be localized in small areas. These areas enlarge as the worms multiply, until the entire pit is filled with worms. It is not necessary to wait until then before cleaning under the cages. But after one or two cleanings, try to avoid taking out the major colonies where the worms have settled and are depositing most of their eggs. There will be some particular areas where you can dig and find simply hundreds of tiny, threadlike worms, and if you look very close, literally thousands of minute, pale yellowish or off-white round capsules about the size of a pinhead. These capsules are earthworm eggs. I called these special areas worm nurseries, and made it a point never to remove the droppings in their immediate vicinity. These areas are usually a few inches to about a foot across, and leaving them undisturbed posed no problems. When the cleaning out was finished, I simply spread a fresh layer of dry sawdust over the entire pit, including undisturbed spots where nurseries were located.

Agricultural lime is another substance which sometimes can be used to advantage. It helps discourage flies from inhabiting the rabbitry. Also, where lime is lacking in the soil, just a thin film of lime over dropping pits helps enrich the garden later when the composted manure is added. When worms are used, the lime should not be spread too thick over the pits; a very thin film is all that is needed. Lime tends to dry and apparently too much repels the worms from an area.

A word of caution to those who would like to utilize earthworms: If chemical insect sprays or powders are used on rabbit droppings under cages or on the compost pile, say goodbye to your earthworms! Even the milder, so-called safe insect sprays will destroy your worms. If insecticides must be used in the rabbitry, you absolutely *must* keep the worm pits unsprayed. To do otherwise is to sacrifice the entire worm population.

Index

Recipes are preceded by a star (*).